D0730691

ORGANIC
MANIFESTO

ORGANIC MANIFESTO

HOW ORGANIC FOOD CAN
HEAL OUR PLANET, FEED THE WORLD,
AND KEEP US SAFE

MARIA RODALE

FOREWORD BY ERIC SCHLOSSER

RODALE

© 2010 by Maria Rodale

First published in hardcover by Rodale Inc. in 2010
This paperback edition published in 2011

Foreword © 2010 by Eric Schlosser

Rodale books may be purchased for business or promotional use or for special sales. For information, please write to: Special Markets Department, Rodale Inc., 733 Third Avenue, New York, NY 10017.

Printed in the United States of America

This book was printed on 100% PCW paper stock.

Book design by Joanna Williams

The Library of Congress has catalogued the hardcover edition as follows:

Rodale, Maria.
    Organic manifesto : how organic food can heal our planet, feed the world, and keep us safe / Maria Rodale.
        p.   cm.
    Includes bibliographical references and index.
    ISBN-13 978-1-60529-485-8 hardcover
    ISBN-10 1-60529-485-3 hardcover
    1. Organic farming—Environmental aspects.   2. Agricultural chemicals— Environmental aspects. I. Title.   II. Title: How organic farming can heal our planet, feed the world, and keep us safe.
S605.5.R62   2010
631.5'84—dc22   2010001224

Paperback ISBN 978-1-60961-136-1

Distributed to the trade by Macmillan

2 4 6 8 10 9 7 5 3 1 paperback

We inspire and enable people to improve their lives and the world around them.
www.rodalebooks.com

To all the leaders of the world,
who hold our future in their hands.

To my whole family who shares this honorable path with me.

And to the farmers.

CONTENTS

FOREWORD

ERIC SCHLOSSER

Pesticides are poisons. They are manufactured to kill insects, rodents, fungi, and weeds. But they can also kill people. Organophosphates—one of the most common types of pesticide—were developed in Nazi Germany to be used as chemical weapons. It was later recognized that the same sort of nerve gases formulated to attack enemy soldiers and civilians could be used against agricultural pests. During the past 60 years organophosphates, organochlorines, N-methyl carbamates, synthetic pyrethroid insecticides, herbicides, fungicides, and fumigants have been applied to the American landscape on a massive scale. During the first year of President George W. Bush's administration, the Environmental Protection Agency (EPA) stopped counting how much pesticide was being used in the United States. The federal government hasn't conducted a survey of pesticide use since 2001. And the chemical companies that sell these poisons don't seem eager to let people know how much is really being sprayed. A conservative estimate of current pesticide use in American agriculture would be about 1.2 billion pounds a year—about 4 pounds of the stuff for every American man, woman, and child.

Despite industry claims that widespread pesticide use
poses little threat to the public health, the latest scientific
evidence suggests otherwise. Almost everyone in the United
States now has pesticide residues in his or her blood. The
effects of direct exposure to various pesticides aren't disputed.
Pesticides can cause damage to the central nervous system,
brain damage, lung damage, cancer, birth defects, sterility, and
death. The long-term effects of minute residues within the
body are more subtle. Recent studies suggest that behavioral
and developmental problems may be linked to childhood
pesticide exposure. And that exposure begins at almost the
moment of conception. Pesticide residues are routinely
detected in the amniotic fluid of pregnant women.

A number of the same companies that unleashed a witch's
brew of toxic chemicals into the environment have, during the
past decade, eagerly promoted the introduction of genetically
modified organisms (GMOs). Monsanto—once the manufacturer of such notorious pesticides as DDT and Agent Orange—
now controls about 90 percent of American soybean
production, through patents on its genetically modified seeds.
The company also controls about 60 percent of corn production. Its leading competitor in the market for GM seeds is Dow
Chemical—a firm that over the years has manufactured
pesticides, plastics, napalm, and the plutonium cores of nuclear
weapons. Although the health risks of eating genetically
modified foods haven't been established, these new transgenic
seeds have been dispersed throughout millions of acres of
American farmland without much study of their potential

effects upon human beings or the environment. The federal panel that approved the speedy introduction of GMOs during the 1990s was headed by Vice President Dan Quayle. A man who couldn't even spell the word "potato" was responsible for allowing the sale of genetically modified potatoes.

The technological mindset that would dump billions of pounds of deadly chemicals onto the soil, and mix the genetic material of different species, and build factory farms where livestock are treated like industrial commodities, and clone animals in order to give them a uniform size, has a deeply arrogant view of the natural world. It regards Nature as something to be conquered and controlled for a short-term profit.

The organic movement embodies a different mindset. It takes the long view. It seeks the kind of profit that can last for generations. It regards the natural world with a profound reverence and humility. It aims to work with Nature—and considers the whole notion of controlling Nature to be absurd. At the heart of the organic movement is a belief in the interconnectedness of things. What you put into the soil winds up in the crops that grow in the soil, winds up in the animals that eat those crops, winds up in the people that eat those animals—and ultimately winds up in everything, since all things return to the soil. As all of us will one day return to the soil.

Maria Rodale is uniquely qualified to speak on behalf of this movement. Her grandfather, J. I. Rodale, founded the magazine *Organic Farming and Gardening* in 1942, just as the chemical-based, industrial agricultural was taking root. Her parents, Robert and Ardath Rodale, devoted their lives to championing organic agriculture and a holistic approach to living. Maria brings to the subject not only her family legacy, but also her personal experiences as a mother, businesswoman, and activist who has been around farmers and farming all of her life. She understands the inextricable link between the health of the land and the health of the society built upon it. This book conveys the importance of the organic movement, at a time when a handful of corporations are trying to control our food supply, when climate change and future petroleum shortages threaten the entire basis of today's industrial agriculture. This book is full of Maria's wisdom, optimism, and common sense. In a remarkably brief period of time, we have done tremendous damage to the environment and to ourselves. Here you will find how we can start to undo it.

INTRODUCTION

Think about what you think you know about organic food for a
moment:

Some people say it's more nutritious, but others say it's not.
Maybe it's the right choice for wealthy developed nations, but is
it practical for underdeveloped and poor countries? Isn't it
more important to eat food produced locally? What's the
difference between organic and natural? And why are organic
foods so expensive? Doesn't the high cost of organic food make
it an elitist indulgence rather than a realistic choice for the
majority of working families? And we seem to hear all the time
that it's not even possible to grow enough food to feed the
world organically. What's the truth?

We are, as a culture, bombarded by so much information all
the time, yet much of it draws conflicting conclusions and never
seems to provide a definitive, clear answer to the question of
why we should opt for organically grown foods rather than those
grown with synthetic chemicals[1]. We live in a fairly constant
state of confusion about most of the things that are essential to
our survival, but have been conditioned to do what is easiest. We
are, frankly, too busy and distracted to think too deeply about
most things. Meanwhile, we eat what's available when we are

hungry and try not to think too much about it because it's all so disheartening and confusing.

Well, I've thought a lot about it. Even more important, I've done the research. In the process of writing and researching this book, I found that there are clear and conclusive scientific data that support my hypothesis that organic agriculture[2] is the key to our survival. The studies are buried deeply within the databases of government agencies, too complicated to be reported by the mainstream media, and subjected to downright suppression at some universities. A quick peek beneath the shiny veneer of the marketing materials of major chemical companies reveals a world where the overriding drive for growth and profits has created a culture of denial that makes victims of us all. I talked with leading scientists and doctors. I read shocking historical accounts of forgotten periods in our past. I spoke with organic farmers, business leaders, and writers. But most important, I spoke with many chemical (or "conventional") farmers from around the country. It was their stories that made it even clearer to me that we need to change.

I am hardly the first in my family to reach the conclusion that we must embrace a return to chemical-free gardening. It's in my DNA. My grandfather, J. I. Rodale, launched *Organic Farming and Gardening* magazine in 1942, and with it the organic movement in America. My father, Robert, also believed in supporting local farmers and buying local foods. He had devoted his life to his father's mission of improving our health and environment through food and how it is grown. They both knew how closely food is entwined with health, and they launched *Prevention* magazine in 1950. To my grandfather and father, organic foods were obviously healthier for people and better for the environment. But in the mid-1960s, they still hadn't proved it to the world. In fact, many so-called experts considered them crazy, or at least highly eccentric. So they set out to prove unequivocally the wisdom of their ideas.

Before my father died in 1990, he began what is now the longest-running scientific study comparing synthetic-chemical versus organic agriculture. He established the Farming Systems Trial in 1981 at the Rodale Institute Research Center in Maxatawny, Pennsylvania. He knew that to prove the viability of organic agriculture, he needed to compile empirical evidence from a scientific setting that compares side-by-side plots of organic and chemically grown foods. He truly loved farmers and farming, and he wanted to see them make a decent, healthy, and independent living. So he bought land, hired a few scientists who were willing to put up with the ridicule of their peers, and made a commitment to keep going for as long as possible—or necessary. Over the years, the government finally did begin to help fund and support the research. Today, university researchers from around the world study our findings, and new organizations have sprung up globally to support and promote organic agriculture.

As a result of my father's and grandfather's efforts, as well as those of a few dedicated and brave scientists willing to buck the overwhelmingly synthetic-chemical-centric system that funds most agricultural research, we now have a conclusive answer to the question of why we should opt for organic farming practices. The Farming Systems Trial shows clearly that organic farming is not only more productive than chemical farming (especially during times of drought and flood), but that soil farmed organically is a key component to solving our climate crisis. But even if you don't believe there is a climate crisis, there are other reasons to embrace organic methods.

In fact, so critical is this imperative to embrace organic methods that I make this statement: If you do just one thing— make one conscious choice—that can change the world, go organic. Buy organic food. Stop using chemicals and start supporting organic farmers. No other single choice you can make to improve the health of your family and the planet will have greater positive repercussions for our future.

This book is about that one thing we can do to save ourselves (because as science has established, the planet will be just fine if we disappear). It may seem hard to believe that one change will make such a big difference, but I will show how it will have a pervasive impact on global climate change while also preserving our health. In fact, it will infiltrate and improve almost every aspect of our lives.

We still have time to heal the planet, feed the world, and keep us all safe. But we have to start right now. To implement this change, we must unite to fight the most important fight of our lives, perhaps one of the most important global struggles in the history of our species. It's a fight for survival. Because no matter what our political beliefs, our religion, our family values, our sexual preferences, our tastes in music or foods are, we are all in this together. Our fates are linked.

We are all one tribe.

This is my organic manifesto.

PART 1

THE GREAT CHEMICAL EXPERIMENT (IN WHICH WE ARE ALL GUINEA PIGS)

We are dealing with 10 global issues at the moment: food security, availability of water, climate change, energy demand, waste disposal, extinction of biodiversity, soil degradation and desertification, poverty, political and ethnic instability, and rapid population increase. The solution to all of these lies in soil management.

—Rattan Lal,
soil scientist, director of Ohio State University Carbon Management and Sequestration Center, January 29, 2009

1. WE HAVE POISONED OUR SOIL, OUR WATER, AND OUR AIR

When I was a little girl, one of my favorite outings was a Sunday drive to the local orchard. Out of the blue my dad would say, "Let's go for a drive," and we'd all scramble into the station wagon. When the destination turned out to be the orchard, we would rush to the juice machine and push our paper cups under the spout to get a cup of cold, fresh cider.

The autumn air would be filled with the scent of fallen leaves and wisps of wood smoke, and we would always come home with a wooden bushel basket full of apples, or sometimes even two if we were going to make applesauce. None of us wore seat belts and, in fact, I doubt the car even had them—it was the 1960s. We headed home feeling lucky to live in a place where apples right from the tree were so delicious, so fresh, and so close by.

These were our good times. My Eden. My family lived right next door to my grandparents on a working farm where chickens, pigs, cows, sheep, and organic vegetables and fruits were raised. The fields were planted with hay and corn. At that time we knew the only way to get organic food was to grow it ourselves and so we did. And it was good.

Years pass, and now I'm driving my own kids to the local fair. In the intervening years, my grandfather died, having achieved iconic status in the hippie culture (although he himself was not one). My father, too, died, killed in a freak car accident while trying to launch an organic gardening magazine in Moscow. As I drive, I notice that the orchard has been turned into a housing development, and only a few gnarled old apple trees remain at the edges of the manicured lawns. I have often joked that Pennsylvania's biggest farm crop is houses, so while I am saddened to see a housing development there, I am not surprised. Nor, unfortunately, am I surprised when I read in the local newspaper that every one of the 800 water wells in that development tested positive for lead and arsenic. The soil is also contaminated, with levels more than 50 times higher than is deemed safe by the Environmental Protection Agency. The families who bought houses on Macintosh or Dumpling Drive, thinking they were getting their slice of the American dream, now are living an American nightmare.[1]

It's impossible not to feel for those poor families. Their children have a much higher risk of suffering reduced intelligence, behavioral problems, and health issues. Consider the young couples who thought they would be starting families and now find themselves unable to conceive,[2] and may even be facing cancer treatments instead of fertility treatments. Consider the hard working people who might never be able to sell their houses. Perhaps you know other families in similar situations.

Now consider this: We are all in the same situation to varying degrees. We are all being poisoned, contaminated, sterilized, and eventually exterminated by the synthetic chemicals we have used for the last 100 years to grow our food and maintain our lawns, to make our lives easier and "cleaner" and our food "cheaper."

Most of us probably think our species' biggest problems, aside from the global economic collapse, have to do with energy and energy independence. The debate over the climate crisis and environmental destruction has been almost completely focused on energy usage—how we drive our cars, heat our homes, and power our affluent and well-lit lifestyles. We haven't yet made the full connection between how we grow our food and the impact it can have on our climate crisis *and* our health crisis.

The global economic upheaval in 2008 and 2009 has afforded us a once-in-a-lifetime opportunity to rebuild and re-envision an economic model, a government, and a future that is based on what is right for people, the environment, *and* business. We can and must create a world that is more than sustainable, that is regenerative. Nature, under optimal circumstances (mainly, when we leave it alone) heals itself. Regeneration is necessary to heal the damage we have already done to ourselves and to our environment.

It is time to begin the process of healing.

WHY CARBON REALLY MATTERS

Over the last century we've all been subjected to an unprecedented chemical experiment. While there have been antivivisection movements around the world to protect animals from testing, I've never heard about a single protest to save our children from this vast experiment. Yet there is increasing and frightening evidence that agricultural and other industrial chemicals are causing significant and lasting health problems—problems that will be hard to solve and take time to correct even if we start making changes today. The evidence is starting to pile up.

Do you know what the number one reason is for kids missing school these days? It's not colds or the normal sicknesses that all children go through during their lives. It's asthma.[3] Asthma's prevalence increased by 75 percent from 1980 to 1994[4] (the last time it was officially measured by the Centers for Disease Control and Prevention). Thirty-four million Americans have been diagnosed with asthma, and worldwide the number has reached approximately 300 million. [5] You could say that it's getting harder and harder to breathe on this planet.

What does asthma have to do with carbon?

Let's think for a minute about the human body and its relationship with the planet. Breathing is fundamental to life. We can live a few weeks without food, a few days without water, forever without cars if we must. But take away our air or our capacity to breathe it in and we are dead in minutes.

We breathe *in* oxygen and breathe *out* carbon dioxide in an ongoing cycle. Researchers have determined that global warming or climate change is caused primarily by *too much* carbon dioxide being produced by cars, manufacturing, and other uses of fossil fuels. Our collective exhaling is exceeding the earth's capacity to process it back into air for us to breathe in.

Carbon, the building block of life,[6] is one of the most abundant naturally occurring elements on earth—it's in coal and inside our bodies, it's in limestone and in every living thing (which is how scientists are able to use carbon dating to determine the age of artifacts and fossils), it's in oil and it's in the air, it's in wood and it's also in soil. In its densest form, carbon is a diamond. The very same element in a less compact form is charcoal or graphite.

Carbon molecules move all the time—and react readily with other elements, especially oxygen. When one carbon atom reacts with one oxygen atom, the result is carbon monoxide, which is both highly toxic and at the same time useful (it's the blue flame burning on your gas stove, for instance). The carbon monoxide reaction occurs most often when carbon is burned in an oxygen-starved environment, like a woodstove. We have all heard stories about people who went to sleep on a cold winter night but never awoke the next morning because their faulty heating systems—oxygen-deprived, carbon-burning combustion—killed them with carbon monoxide.

When one carbon atom merges with *two* oxygen atoms, the result is carbon dioxide (CO_2). Carbon dioxide is produced naturally by things like volcanoes and hot springs, but it also occurs when you burn carbon. Like carbon monoxide, carbon dioxide is toxic in high concentrations and can cause dizziness, headaches, rapid breathing, confusion, palpitations, and at very high concentrations death.[7]

Oxygen is released into the air by plants through photosynthesis. Plants breathe *in* carbon dioxide and breathe *out* oxygen, a fundamental reason that plants of all kinds are essential to our survival. Plants generate oxygen we need to survive. The earth doesn't have enough plants to breathe in and store all the carbon dioxide our activities have been producing and recycle it as fresh oxygen. So we either have to stop spewing out so much carbon dioxide or find ways to "sequester" it—hold it someplace. This is the conundrum we now face. This is the essence of our climate crisis.

There is no shortage of schemes and dreams to solve this problem—including making "biochar,"[8] cap-and-trade programs,[9] creating vast underground tanks to hold the carbon, or shooting it out of our atmosphere and into space.

But what if we are missing a major piece of the equation? Most discussion on global warming has focused on the energy issue, both because it's the most visible cause of carbon dioxide emissions and, more important, because it's where all the money and political power are concentrated. Oil, gasoline, "clean" coal, solar, wind, biofuel, and all that goes with those things (wars, power grids, automobile companies, bailouts, deals, lobbying, government appointments) have been hogging our attention. And so in our daily confusion, we just take for granted that we will always have food, comfortable lifestyles, cars, and climate-controlled homes.

We take it all for granted—just like breathing.

Now imagine for a minute that someone, maybe Bill Gates, has developed a nanotechnology for sequestering carbon (that is, taking the excess carbon dioxide that causes global warming from the air and holding it in a stable, safe form somewhere where it cannot do any damage to the atmosphere). Perhaps it is a technology that you put in the soil that will suck up all the carbon we have expelled into the air. Bill will set up a new business called Mycrosoft that is backed by venture capitalists and has an IPO scheduled for Year 2. People would be all over this like girls at a Jonas Brothers concert. Headlines in the *New York Times* and the *Wall Street Journal* would herald this new technology as the savior of our environment. Warren Buffett would buy a piece of the business, for sure. Bill Gates would win a Nobel Prize. All the lucky investors would make a fortune.

The irony is that this cutting-edge, breakthrough technology already exists. It's just that nobody has figured out how to own it yet.

(Actually, it's just a matter of time until Monsanto figures out a way to make money from it. Monsanto has been hosting meetings of the Agricultural Carbon Sequestration Standard Committee, along with the USDA, to develop standards for "validating carbon offsets resulting from soil carbon sequestration of greenhouse gas emission reductions at the soil interface." In other words, it is trying to figure out how to make a business out of carbon sequestration in the soil. This process is facilitated by a company called Novecta, a joint venture of the Iowa and Illinois Corn Growers Associations that provides guidance on crop protection, "value enhanced crops," and help farmers understand their role in "producing product that is desired by the food, fuel, and industrial markets."[10]

These are smart people who are on to carbon sequestration and trying to get ahead of the market, the government, and the organic community in order to control it and make money from it. And they have billions of dollars at stake in doing so.)

What is this magical, superpowered natural nanotechnology? Mycorrhizal fungi.

"Myco" means fungus, and "rrhizal" means roots. So Mycorrhizal fungi are literally fungi that grow on the roots of plants.

For more than two decades the Farming Systems Trial (FST) at the Rodale Institute has been studying what happens over time to plants and soil in both organic and synthetic-chemical farming systems. The most surprising finding of all has been that **organically farmed soil stores carbon.** A *lot* of carbon. So much, in fact, that if all the cultivated land in the world were farmed organically it would *immediately* reduce our climate crisis significantly. "These fungi actually build our soil and its health and contribute to taking greenhouse gases out of the air—counteracting global warming to boot," says Paul Hepperly, PhD, a Fulbright scholar and former senior scientist at the Rodale Institute.

Conversely, soil farmed using synthetic-chemical or "conventional" methods has very little ability to keep or build vital supplies of carbon in the soil. This is not surprising, since farmers often apply fungicides as well as chemical fertilizers and pesticides. These chemicals are meant to kill. As a result of using these chemicals, a farmer is left with debilitated soil that has weakened microbial life, a compromised structure, and a significantly impaired ability to withstand the stresses of drought and flood.

The fact that we haven't noticed these little helpful creatures before shouldn't surprise us. We prefer our nature in the macro—the postcard vistas and views. When it comes to the micro, we'd rather not look or know. We know more about outer space than we do about the ground we live on, about the soil that sustains us. In general we don't care to think too much about soil. Frankly, it's not sexy. In its verb form, it's a synonym for something that's dirty or ruined. Our most regular contact with soil probably occurs when it gets tracked into the house on muddy shoes. Then we get out the bucket and mop and fill it with fresh meadow-scented antibacterial cleaners to purify our homes and protect our families.

In the 1950s, a promotional brochure for DuPont Farm Chemicals trumpeted "Man against the soil: The story of man in his rise from savagery to civilization is the story of his struggle to wrest his food from the soil." Soil is our enemy?

Television commercials for cleaning products show magnified images of little bacterial villains who are out to get us, making us paranoid and afraid. And yet, according to Lynn Bry, MD, PhD, clinical fellow of pathology at Brigham and Women's Hospital at Harvard Medical School, if all the germs and bacteria in our bodies (and all around us) were eliminated, we would be dead within 2 weeks.[11] Why, then, are we so intent on wiping them out?

Suspend your fear of dirt and all those things we can't see with our own eyes for a minute.

What we call "soil" is a living thing. Just one tablespoon of soil can contain up to 10 billion microbes—that's one and a half times the total human population. We are learning more each day about what goes on in that soil. The discoveries are surprising—and incredibly important.

Right now, soil scientists understand less than 1 percent of all the living things in the soil. But soil is more like us than like plants because the microbes in it breathe in oxygen and breathe out carbon dioxide. Healthy, organic soil also stores massive quantities of carbon and holds it tightly, just like a tree holds on to and stores carbon in its trunk and limbs (which is why all of our forests, including the rain forests, are so important to our survival).

Think about this for a minute.

It's about more than our climate crisis. It's about more than whether or not we can feed all of the people on the planet. When we destroy the normal functioning of our soil and spew more carbon dioxide into the air than we can sequester, **we are suffocating ourselves.** We interrupt the essential balance between people and the planet, people and nature, and soil and plants. It's much, much graver than melting icebergs and rising ocean levels. It's graver than food riots and graver than dead zones the size of Texas in the Gulf of Mexico. All that excess carbon dioxide is like a giant pillow being pressed down slowly but surely over our faces, and just as surely the earth is losing the ability to fight back on our behalf.

HOW WATER ALSO
REFLECTS OUR MISTAKES

Unfortunately, it's not just the soil and air that have been affected by our chemical contamination. Currently, 60 percent of the fresh water in the United States is used for agricultural purposes.[12] And when it's used for chemical agriculture, which is by far the majority, all those chemicals leach through the soil and into the waterways and wells to poison our drinking water, our rivers and streams, our bays and oceans, and, ultimately, all of us. Agricultural chemicals currently account for approximately two-thirds of all water pollution. According to an article in the journal *Science*, marine "dead zones"—areas in which fish, plankton, crustaceans, and other ocean life cannot survive— have been doubling in size every 10 years since 1960. Sea life in these dead zones experiences hypoxia literally **suffocating** because the water is starved of oxygen.[13]

Water is the ultimate recycled product. It rains, the plants drink the water, the soil cleans the water, people dig wells and drink the water, and it rains again. The water we drink every day (even if it purports to come from Fiji or the Evian Mountains in France) has followed this cycle through the earth, humans and animals, and plants an infinite number of times. We rely on natural processes to clean the water. But neither nature nor our high-tech water filters can remove all the toxic chemicals from water. They build up and linger for a long, long time. And they have the potential to poison us. It has been estimated that it would take an *immediate* 45 percent reduction

in the amount of agricultural chemicals applied to our soils to
have any impact at all on slowing the growth of the dead zones
in our coastal waters.[14] Without major global governmental
involvement and an outright ban on chemicals however, a 45
percent reduction is unlikely.

An increasing source of concern regarding water is the
drugs we use (and often overuse), many of which are made by
the same companies that make agricultural chemicals. We are
trapped in a cycle I call the "chemical death spiral." The majority of research funding is being spent on the quest for miracle
cures rather than preventing disease. Yet in many cases the
same people who are benefiting from selling those "cures" are
also contributing to causing those diseases in the first place.
Add to that the human impulse to pay for a cure rather than
change our unhealthy behaviors and you have a recipe for
capitalist success and built-in market share growth that will
expand until it all collapses in on itself and stops working.

Arsenic is a prime example. It is still used extensively in
farming as a pesticide. The Environmental Protection Agency
(EPA) has set the safe limit in drinking water wells at 10 parts
per billion (ppb), but in many areas around the United States,
levels range from 50 to 90 ppb. In some Asian countries, the
levels exceed 3,000 ppb. In one recent study published in
Environmental Health Perspectives, mice given water containing
100 ppb of arsenic had much more serious upper respiratory
symptoms when exposed to the swine flu virus than those
mice who had clean drinking water. The same study linked
chronic low-level exposure to arsenic to cancer, cardiovascular

disease, diabetes, and reproductive and developmental defects.[15] It was also the poison of choice for murderers in the days before forensic science became available to the police.

Going back to asthma, there is no cause widely accepted by the medical community. Asthma killed 3,613 people in 2006.[16] Many studies show a link between urban living, chemicals, pollution, and other environmental factors. The Agricultural Health Study, funded by the National Institutes of Health, has been studying agriculatural workers' health issues in Iowa and North Carolina since 1994. A 2008 newsletter to participants points out a need for more in-depth study on lung health of farmers because "research shows that farmers and their families may be more likely than the general population to have asthma and other respiratory problems." The article notes, interestingly, that in general, women who grow up on farms are less likely to have asthma than women who don't grow up on farms. However, "if they applied chemicals, they report more allergic asthma than others in the group."[17]

Meanwhile asthma rates increase as the synthetic-chemical agriculture that destroys the soil's ability to function proliferates globally. And it gets harder and harder to breathe.

THE LAND HAS RUN OUT

For thousands of years we've had the luxury of farming, grazing, deforesting, and degrading the land until it was exhausted and turned into desert, and then moving on to new land.

We no longer have that luxury.

Millions of acres around the globe have been destroyed by deforestation, desertification, and destructive farming practices, and if we continue on our current path, there is no end in sight to the devastation.

Now we need to learn how to *stay*. We need to figure out how to grow our food in ways that regenerate the land, rather than destroy it. We need to find ways to express our natural desires for growth, wealth, and creation without ruining the very sources of our strength and power.

The *desire* to stay is real. We love our homes, our land, and our countries. We connect with our communities and our shared histories. We love our families and want them to survive. And yet our behaviors often undermine our desires.

Our lands *seemed* limitless, but now we know we have reached the end of land and most of it is already contaminated and destroyed. A subprime farming crisis is looming on the horizon, yet the majority of Americans still believe that the only way we can feed the world is with synthetic chemistry, biotechnology, and other artificial means.

There's just one problem. The chemical system of agriculture is killing us.

2. WE HAVE POISONED OURSELVES AND OUR CHILDREN

We have been conditioned to fear many things, especially what is different from us. We are afraid of bugs, bacteria, and things that are scary, so we kill them. We have yet to make the connection that in killing other things we are also killing ourselves.

The environmental impact of agricultural chemicals alone is, to my mind, sufficient scientific and ethical argument for putting an end to their use. But agricultural chemicals such as pesticides, fertilizers, herbicides, and their household counterparts are destroying more than soil and water.

Like many people, I am worried about our climate crisis. But in certain locations, you can imagine a future where global warming is not that bad. It might mean warm weather all the time, new beachfront properties, and growing citrus trees or coffee plants in Pennsylvania. And if you are truly honest with yourself, saving the polar bear might not make your top 10 list of things to do this year. It is heartbreaking to watch mother polar bears and their babies drowning because they can't swim far enough to reach ice or land. But it is so far away and feels so hard to address. Maybe you write a check and feel better, but that doesn't really change much.

Meanwhile, autism and attention-deficit/hyperactivity disorder (ADHD), diseases virtually unheard of a few decades ago, are now diagnosed regularly. Of every 100 children born today, one will be diagnosed with autism before the age of 8.[1] About 4.4 million children between the ages of 4 and 17 have been diagnosed with ADHD. Rates of asthma, diabetes, and childhood obesity are at all-time highs and scientists can't explain why the number of children with food allergies has increased 18 percent in the last decade.[2] Is it a coincidence that the prevalence of these problems has increased as we have increased the use of chemicals to grow our food?

Experts might claim that our reporting and diagnostic technologies are better than they used to be, which is probably true. But compared to other countries where the reporting is just as good (if not better), we in the United States spend far more on health care, but have dismal results. Our life expectancy is the shortest and our infant mortality rate is the highest of any developed nation.[3] In many of the countries whose citizens have longer life spans than Americans do, a lot of the chemicals that we believe are necessary to grow food have already been banned.

On his way out of office, President George W. Bush halted the program that tests pesticide levels on fruits, vegetables, and field crops because the cost—$8 million a year—is "too expensive."[4] That's just one small example of where our priorities are when it comes to protecting our health and that of our children.

According to the Mount Sinai Medical Center Children's Environmental Health Center (CEHC) in New York City, more than 80,000 new chemical compounds have been introduced since World War II. Many of them are used in agriculture. There are 3,000 so-called high-production-volume chemicals, meaning that more than 1 million pounds of each are produced or imported in the United States each year. More than 2.5 billion tons of these chemicals are released into the environment in the United States alone *each year*. In addition, more than 4 billion pounds of pesticides are used annually in the United States—to kill everything from agricultural pests to inner-city cockroaches to microbes and bacteria in schools and hospitals. Traces of all of these chemicals can be detected in virtually each and every one of us. Yet only half of the compounds have been even minimally tested and less than 20 percent have been tested for their effects on fetal nervous systems. (What parent would *agree* to that sort of testing in the first place? And yet all of our children have already been exposed.) At least 75 percent of the manufactured chemical compounds that *have* been tested are known to cause cancer and are toxic to the human brain.[5]

"Failure to test chemicals for toxicity represents a grave lapse of stewardship," says Philip Landrigan, MD, professor and chairman of the CEHC. "It reflects a combination of industry's unwillingness to take responsibility for the products they produce coupled with failure of the US government to require toxicity testing of chemicals in commerce."

The Harvard-trained Landrigan is a pediatrician and epidemiologist who has dedicated his life to being a leader in public health and preventive medicine. He's worked for the Centers for Disease Control and Prevention and the National Institute for Occupational Safety and Health and has published more than 500 scientific papers (as well as five books). A retired captain in the US Naval Reserve's Medical Corps, Dr. Landrigan chaired the National Academy of Sciences group whose research led to the passage of the 1996 Food Quality Protection Act, which is the only environmental law with explicit provisions for the protection of children. You could not find a physician or activist who has been more committed to protecting children's health from environmental toxins. His pioneering research into the toxicity of lead at low levels was responsible for the federal mandate to remove lead from gasoline and paint. (Lead arsenate was also heavily used as an agricultural pesticide before the introduction of newer chemical compounds.) Dr. Landrigan has also worked hard with some success to get other chemicals banned from homes. These include organophosphates, which are toxic to the brain and nervous system and cause thousands of deaths from poisoning every year. Unfortunately, they are still commonly used as pesticides on farms today and residues are often tracked into homes.

A study published in 2009 in the journal *Therapeutic Drug Monitoring* documented the first *definitive* connection between the most common childhood cancer, a form of leukemia, and the very organophosphate pesticides that Dr. Landrigan has been trying to get banned.[6]

Dr. Landrigan is also concerned about the widespread use of antibiotics in raising animals for human consumption. The devastating practices employed in large-scale meat production have been the subject of entire books. Dr. Landrigan characterizes these facilities as "animal slums." Animals raised in those conditions frequently suffer from stress-related infections, so they are routinely treated with antibiotics as a "preventive" measure. Even worse, antibiotics are fed to livestock to *promote growth* (since it's cheaper than using real food). The rampant use of antibiotics to treat all food animals (cattle, hogs, and poultry) in these slums is shortsighted and dangerous, as many doctors have attested. Even the FDA is calling for a ban.[7] Could you imagine keeping your children on antibiotics full-time just to keep them "healthy"?

CHEAP FOOD EQUALS
HIGH HEALTH CARE COSTS

To feed our demand for cheap food, we have put ourselves and especially our children's lives at risk. According to an article in the *Archives of Otolaryngology*, there has been an "alarming" increase in drug-resistant infections in children—especially in the ears, sinuses, head, and throat (tonsils). In the period between 2001 and 2006, MRSA (methicillin-resistant *Staphylococcus aureus*) head and neck infections caused by drug-resistant bacteria in children have more than doubled—from

12 percent in 2001 to 28 percent in 2006.[8] Dr. Landrigan
attributes this increasing resistance to overuse of antibiotics in
raising our food. When asked if he sees increased incidence of
drug-resistant infections in his own practice in New York City,
he replies that it is "all over the place" and it "worries the hell
out of me."

Dr. Landrigan has good reason to be worried. MRSA affects
children and healthy adults. It had been confined to hospitals,
but now you can catch MRSA at gyms, schools, day care centers,
and military barracks. The Union of Concerned Scientists has
repeatedly warned of a link between MRSA and the overuse of
antibiotics on large hog farms. A peer-reviewed study in
Applied and Environmental Microbiology confirmed that MRSA
is found on almost half the pork and 20 percent of beef samples
taken from a sampling of 30 supermarkets in Louisiana.[9] The
European Union, South Korea, and many other countries
prohibit preventive use of antibiotics in the factories where
animals are raised for meat.[10] Not in the United States. In fact,
farmers can just go to their local farm stores to buy a 50-pound
bag of antibiotics—without a prescription.[11]

In response to the spiraling rise of bacterial and viral
infections, many concerned parents, hospitals, and schools
have begun indiscriminately using antibacterial products. But
those very products are part of the problem. Think of it this
way: *You and your kids are washing your hands in pesticides.* The
EPA recently allowed triclosan, the predominant antibacterial
agent used in products from shampoo to water bottles to crib
liners, to be re-registered (chemicals need to go through a

periodic re-registration process in which all of their risks are reviewed), even though the EPA acknowledges that triclosan interacts with androgen and estrogen receptors and affects the thyroid gland in rats. The EPA acknowledges that triclosan is linked to antibiotic resistance, and that it is showing up in fish and drinking water. The EPA doesn't have to look at triclosan again until 2013.[12]

The Endocrine Society was founded in 1916 to research the role of hormones on our health. It recently released a major report to raise concern about the effects of chemicals, including the organochlorinated pesticides, on our health. "In this first Scientific Statement of The Endocrine Society, we present the evidence that endocrine disrupters have effects on male and female reproduction, breast development and cancer, prostate cancer, neuroendocrinology, thyroid, metabolism and obesity, and cardiovascular endocrinology. Results from animal models, human clinical observations, and epidemiological studies converge to implicate EDCs [endocrine-disrupting chemicals] as a significant concern to public health."[13] The Endocrine Society also validated what is known as the low-dose or inverse dose response factor, which means that *any* dose—even small ones—have the potential to do major damage. The statement also called attention to the direct link that the Agricultural Health Study has found between increased prostate cancer rates, methyl bromide (a fungicide used heavily on strawberries, among other crops), and many pesticides.

While "endocrine" seems like a fairly nonthreatening term, there is one endocrine disease we are all familiar with:

diabetes. Endocrine disrupters include the organophosphate pesticides, atrazine, bisphenol A, lead, mercury, and many other common chemicals we ingest daily—though we don't intend to.

Dr. Simon Baron-Cohen has a hypothesis about autism that's called "the extreme male theory," which theorizes that a hormonal imbalance leads to "overmasculinization" of a child's brain. Harvey Karp, MD, a world-renowned pediatrician, recently proposed that exposure to endocrine disrupters might be the cause.[14]

Diabetes and autism are both increasing to epidemic proportions. Until recently, synthetic chemicals have escaped blame.

What about cancer? Despite investing billions and billions of dollars in research money, cancer death rates have remained fairly flat since the 1950s.[15] Two well-researched books, *The Secret History of the War on Cancer* by Devra Davis, PhD, and *The Politics of Cancer Revisited* by Samuel S. Epstein, MD, reveal that experts have known since the 1930s about the connection between environmental toxins, hormones, and cancer. And yet, as Dr. Davis documents in her book, the companies that fund lobbying groups have actively suppressed information, infiltrated and run the charitable organizations that are supposed to cure them, and invested in huge advertising campaigns to create a public sense of confidence in the very companies that are withholding and perverting the truth.

"In America and England, one out of every two men and one out of every three women will develop cancer in their lifetime,"

Dr. Davis writes. "Cancer is the leading killer of middle-aged persons, and, after accidents, the second-leading killer of children."[16] Both she and Dr. Epstein list as irrefutable causes of cancer pesticides, including atrazine and arsenic; hormones, including artificial growth hormones that are used on animals; and the thousands of chemicals (and plastics) that are hormone disrupters. It's truly an outrage that our government has not done more to prevent the use of these known carcinogens. But, as I will show you in greater detail, researchers in our hospitals and universities are often funded by the companies who produce the chemicals being studied. The research, unless clearly supportive of the funders and their products, is often suppressed or simply ignored.

Over the years, the Agricultural Health Study (AHS) has established a definitive link between pesticides and Parkinson's disease, certain cancers and diabetes.[17] The following excerpt from the study findings, as reported in the National Institutes of Health newsletter, shows just how much we already know about the toxic effects of pesticides, and just how little we are really doing about it. (The emphasis on phrases below is my own.)

Research involving pesticide applicators in the AHS shows that exposure to some agricultural chemicals may increase the risk of diabetes, *confirming the findings from earlier studies.*

The study found a link between diabetes and seven pesticides: aldrin, chlordane, heptachlor, dichlorvos, trichlorfon, alachlor and cyanazine. The strongest association with the disease was found for trichlorfon, although the number of applicators with heavy use was small.

Scientists with the National Institute of Environmental Health Sciences (NIEHS) analyzed data from nearly 1,200 participants in North Carolina and Iowa who developed diabetes after they enrolled in the long-term AHS study.

"The burden of diabetes is increasing around the world," said Dr. Dale Sandler, who oversaw the research at NIEHS. "We hope what we've found will inspire other scientists to pursue additional studies on this important issue."

Although three of the insecticides studied—chlordane, aldrin, and heptachlor—are no longer on the market, measurable levels of these and other pollutants are still detectable in the general population and in food products. These chemicals are organochlorines, as is dioxin, which has been shown to increase the risk of diabetes among Vietnam War veterans exposed to Agent Orange.

Participants who had used the herbicides alachlor and cyanazine had a higher risk for developing diabetes, particularly those participants who had used these chemicals repeatedly over their lifetime.

"Because few studies have looked at the association between herbicides and diabetes, *more research is needed to confirm these findings,*" Dr. Sandler said.

As in other studies, AHS results confirmed the known link between obesity and diabetes. In fact, the strongest associations were found among overweight and obese participants. This may be because people with more body fat are more likely to store high levels of pollutants than people who are lean."[18]

This story also illustrates the classic technique of delay and doubt: the perpetual need for "more research," even though this study confirms other earlier studies. Our government pays for research and then doesn't act on it. Continuing to do scientific research is essential. But not acting on it is dangerous.

EVIDENCE IN THE ANIMAL WORLD

There is more disturbing news from the halls of research labs around the world. Researchers are starting to see extremely disturbing trends in fertility rates, gender, and general resilience of certain species (including ours).

It started with frogs and alligators. Why are alligators' penises shrinking? Why are boy toads turning into girl toads? Why are fireflies disappearing from the summer night skies?[19] What's the cause of what has been called colony collapse disorder, which has been causing honeybees to disappear? Why are bats succumbing to the mysterious white-nose disease? Perhaps you wouldn't care if all the bees and bats died—after all, bees can be a pain during summer picnics and bats can be scary. But aside from the fact that these animals are essential to our survival on this planet, they are also indicators of the viability of life for *us*. They are our canaries in the coal mine.

Already we are seeing these indicators infiltrate our own species. Genital deformities in human males are on the rise, as are sperm abnormalities and testicular cancers. And girls are experiencing an earlier onset of puberty and problems that lead to infertility and cancer later in life.[20]

Warren Porter, PhD, a zoology and environmental toxicology professor at the University of Wisconsin, Madison, points to recent scientific findings and concludes that not only are we threatening our own and our children's fertility, but low-level exposures could also pose threats to our immune, hormonal, and neurological systems and especially our developmental processes. These functional changes may be the result of alterations in the regulation and expression of our genes (DNA). What's really scary is that the genetic expression changes may be becoming heritable.[21] Chemicals that have already caused genetic expression changes in rats include vinclozolin, a common fungicide used on grapes, and bisphenol A, found in plastic baby bottles and toys. There is no guarantee that, if environmental chemicals can induce such changes and make them heritable in humans, even if we stop using the chemicals right now, we could return our DNA to a healthy, normal state.

Dr. Porter is no stranger to the chemical companies or the EPA. Many years ago, despite threats from chemical companies, he and his colleagues published a study showing that aldicarb (an N-methyl carbamate), an insecticide commonly applied to citrus, cotton, potatoes, and watermelon crops and added to irrigation water, is a powerful immune suppressant. That means it reduces your body's ability to fight off disease. The greatest effects were at the lowest doses (1 part per billion), 100 times lower than the EPA's safety standard.[22] Rather than remove aldicarb from the market as a result of that study, the EPA ceased funding Dr. Porter's research. Aldicarb is still used heavily today. In fact, it was the chemical that in 1984 caused the disaster at Bhopal, India, killing thousands of Indians instantly and leaving more than 100,000 chronically ill, deformed, and in pain. (Check out http:// bhopal.org for the full horror story.) But you don't need a manufacturing plant explosion to cause damage. Scientists, including Dr. Porter and Dr. Landrigan, are all seeing the inverse dose response to many chemicals—from aldicarb to lead. They have found that in some cases *smaller doses actually do* **more harm** *than larger doses.*[23] This flies in the face of all the government standards that set "allowable safe limits" to chemical exposures. In a scientific paper published in 2003, Wayne Welshons and colleagues demonstrated that at low concentrations of estrogen-mimicking chemicals, the EPA model for assessing biological effects *underestimated* those effects by a factor of 10,000![24] There are no safe limits, no matter how small. As Dr. Landrigan says about lead, "the biggest bang for the buck still occurs at the lowest doses."

Larger doses, of course, cause death.

In 2002, Porter and colleagues published a paper showing that a weed killer commonly used on lawns that includes 2,4-D caused abortions and absorption of fetuses. A representative from a chemical company that sells this product approached a dean at his university and demanded that the paper be retracted. The dean replied that the peer-review process would settle any concerns and, thankfully, responsible scientific process prevailed since there was no credible scientific evidence presented to counter the data in their paper. Yet products that contain 2,4-D remain the most commonly used herbicides in the world today, with more than 46 million pounds applied in the United States every year.[25] Once again, just because a toxic chemical is scientifically proven to be harmful doesn't guarantee that our government will respond in a way that protects us.

Porter's 1999 study shows how atrazine and nitrates in drinking water can alter the aggression levels, thyroid hormone levels, and immune systems of mice. It's not good. Many studies have shown that atrazine, one of the herbicides most commonly used in the United States, is an endocrine disruptor, which has demasculinized frogs, caused mutations in the frogs' testes and ovaries, diminished their essential ability to call for a mate, damaged sperm quality in frogs and possibly humans, and contaminated groundwater.[26] The European Union banned atrazine in 2003, yet more than 76 million pounds are applied in the United States each year. In 2006, the EPA issued a statement declaring that the use of atrazine posed no threat to the US population, including children and infants.[27] It based this conclusion on the results of a few studies done by Syngenta, the maker of atrazine. Although

Syngenta is a Swiss company, it has a very large US business.
(In 2005, the Natural Resources Defense Council, a nonprofit
advocacy group, sued the EPA over what it called "backroom
deals with pesticide makers" like Syngenta.[28]) When you look
up atrazine on the Office of Ground Water and Drinking Water
section of the EPA's Web site, it attributes these health effects
to short-term exposure to it: "congestion of heart, lungs and
kidneys; low blood pressure; muscle spasms; weight loss;
damage to adrenal glands." Long-term exposure can cause
cancer. Thousands of wells in America are contaminated with
atrazine.[29]

Atrazine is banned in Switzerland, which is where Syngenta
is based.

Fortunately, the EPA—under new leadership—has agreed
to review the safety of atrazine.[30] The USDA has stated that
banning atrazine would reduce crop production by 1 percent.[31]

ARE WE ALL ROUNDUP READY NOW?

Before we move on from the topic of chemicals and their
devastating effects on our health, there is one more chemical—
or, rather, combination of chemicals—we need to discuss. Its
brand name is Roundup.

Roundup is the most widely used weed killer in America,
thanks to the intensive efforts of its manufacturer, Monsanto,
to market it. Weeds are both a real problem for farmers and
an aesthetic one. In the fields, weeds compete for energy
with crops, but a weedy field also looks messy (which, believe

it or not, truly matters to a lot of farmers). There are many
organic solutions for weed control, including the planting
of cover crops, mulches, burning, and good old manual labor,
but the easiest way to get rid of weeds is to spray them with
Roundup. Roundup is a broad-spectrum herbicide—which
means that it kills many plants, including crops. For decades,
farmers used Roundup to clear the fields of weeds before
planting, then as a spot herbicide thereafter (relying more
on tilling the soil to control weeds than on spraying). Monsanto then developed corn and soybean seeds altered into
genetically modified organisms (GMOs)[32] to resist being
killed by applications of Roundup. Since 1998,[33] when
"Roundup Ready" GMO seeds were first introduced, growth
has been both controversial and fast. Currently, 91 percent
of all soybeans, 85 percent of all corn, and 88 percent of all
cotton in the United States are grown from GMO seeds.[34]

These plants are exposed to heavy applications of the
herbicide and survive—all the way to our tables. Like many of
the toxic chemicals I write about in this book, Roundup is used
by home owners, too. You'll see commercials for Roundup on
television during spring and summer sporting events, being
promoted as the manly way to get rid of weeds (no squatting
required). There are other herbicides with the same active
ingredient.[35] It's used on farms, golf courses, home lawns,
roadway berms, and railway corridors. Hunters use it in the
woods and city dwellers use it in sidewalk cracks.

Glyphosate is the active ingredient in Roundup, which also
contains surfactants and nonionic (fat-soluble) solvents.

Glyphosate has always been promoted as being fairly inert. And surfactants are purportedly no big deal—they are just the means for helping the glyphosate stick to the plants. I can't tell you how many farmers, foresters, hunters, and gardeners I have talked to who have said, "Yeah, Roundup is totally safe. It breaks down after a few months and we've been told it's no problem." Well, the scientific literature says that that is not true.[36] And because the surfactant is considered inert, the manufacturer doesn't have to tell you what's in it.[37]

But the surfactant allows Roundup to get *inside* the plants that we eat. You can't wash off the contaminants. Roundup is "*in* the plant, not just *on* the plant," Dr. Porter explains. "Fat-soluble chemicals [in Roundup] have the master entry key into the plant and into our bodies, because every cell in our body is a fatty membrane. So anything that is fat soluble can cross the blood-brain barrier and also the placental barrier. Fat-soluble chemicals are the nonionic solvents in these chemical mixtures. In order for any kind of pesticide to have a biological effect on a plant, it's got to get inside a cell to kill it.

"Before Roundup Ready soybeans were on the market," he continues, "the tolerance for Roundup was 3 ppm [parts per million]. Soybean seeds were meeting that requirement, since farmers were not using excessive amounts of Roundup on their crops. Once the Roundup Ready soybeans showed up at the marketplace, they had concentrations up to 20 ppm, indicating that farmers upped the application rate since it wouldn't kill the plants. So Monsanto went to the EPA and asked to have the tolerance raised. Otherwise, the soybeans could not be sold.

The tolerance was raised not only in the US, but in Australia and other countries where substantial amounts of the Roundup Ready soybeans were being grown. (But not in the European Union, which has still banned GMOs) This is why genetically modified crops will have higher concentrations of pesticides in them."

You might be thinking this isn't such a big deal if you don't eat that much soy anyway. Just the occasional edamame serving or soy milk in your latte, right? Wrong. Soy is in everything, from Crisco to infant formula, nondairy creamers to vegetable oil, ketchup to crackers, crayons (!) to veggie burgers and vegan cheese. So if you are a vegan or vegetarian who is not eating primarily organic foods, you and your family could be eating *a lot* of contaminated GMO foods. And, thanks to our government, none of those products need to be labeled as containing GMOs. (But given the extent of their usage by farmers, you can pretty much assume they are in everything not labeled organic or non-GMO.)

Just to be clear: According to Dr. Porter, GMO foods might contain higher levels of chemicals *inside* of them than organic foods.

And they're getting inside of us.

Fortunately, some people are brave enough to keep an eye out for us. A recent report by the American Academy of Environmental Medicine has called for an immediate "moratorium on genetically modified food." The organization, made up of doctors and other professionals interested in the "clinical aspects of humans and their environment," has stated that "several animal studies indicate serious health risks associated

with GM food consumption including infertility, immune
dysregulation, accelerated aging, dysregulation of genes
associated with cholesterol synthesis, insulin regulation, cell
signaling, and protein formation, and changes in the liver,
kidney, spleen and gastrointestinal system. There is more than
a casual association between GM foods and adverse health
effects. There is causation. . . . Also, because of the mounting
data, it is biologically plausible for Genetically Modified Foods
to cause adverse health effects in humans."[38]

The group's goal is to alert and educate doctors, who
are likely noticing increases in strange symptoms and perhaps
not understanding that the food people eat every day might be
causing them.

THE BIG DISCONNECT

What is wrong with us? Why do we seem to care so little about
our own safety, our own health, and the future of our children?
Why are we willing to pay thousands of dollars for in vitro
fertility treatments when we can't conceive, but not a few extra
dollars for the organic foods that might help to preserve the
reproductive health of our own and future generations?

Newspaper editorials and TV programs question whether
or not organic foods are healthier for us or worth the extra
cost, yet they *ignore* the growing concerns of doctors and
scientists about the long-term impact of consuming foods
treated with chemical fertilizers and pesticides. Whether or
not organic foods are more nutritious (some studies have

shown they are, some have shown they aren't) isn't the most important point.

The most important point is that growing scientific evidence suggests that the toxic chemicals we are using to grow food *are* destroying us.

Plant nutrition is a reflection of the ongoing degradation of our soil quality, seeds, and farming methods. The USDA and scientists elsewhere have been measuring the nutritional value of different foods for more than 50 years and have found significant nutrient declines in *all* crops in *all* regions over the past several decades.[39] Scientists disagree on why this is happening, suggesting everything from inconsistent methods of measurement to agribusiness's relentless quest for higher yields, but the USDA to date has shown a shocking lack of interest in the problem. And government funding for nutritional research is microscopic compared to funding for other types of research.

Here is what we do know: Plants, animals, and people have immune systems. When we are in a natural environment and take good care of ourselves—eat right, exercise, sleep well, use basic hygiene techniques, feel loved and cared for, and actively take part in our communities—our immune systems are typically healthy and strong. Constant exposure to natural pests taps into our inner resilience and makes us stronger, enabling us to develop antibodies to fight external threats. But when we try to sterilize our environment (or the other extreme, let it be truly dirty) or try to exterminate a weed, an insect, or a disease, nature fights back just as we would, launching even stronger attacks. The result is viruses, diseases, and superpests that become resistant to pesticides. *The more we try to isolate ourselves and control nature, the weaker and more vulnerable we become.*

As Bill Miller, MD, chair of the Department of Family Medicine at the Lehigh Valley Health Network in Allentown, Pennsylvania, has observed, as a rule we are "overfed and under-germed."

There is so much more we can study to help illuminate the problems with chemicals. But there is enough evidence to know now that synthetic chemicals are destroying our health and our ability to reproduce and, thus, our ability to survive as a species. Agricultural chemicals have statistically and significantly been implicated in causing all sorts of cancers, behavioral problems, attention-deficit/hyperactivity disorder, autism, Parkinson's disease, reduced intelligence, infertility, miscarriage, diabetes, infant deformities, and low birth weight. And with endocrine disruptions come genital deformities, early puberty, gender "issues," and, again, diabetes and cancer. But all this research comes from the few scientists courageous enough to swim against the tide, to resist the easy funding offered by chemical and pharmaceutical companies and the pressure of their peers who rely on that funding.

We are allowing a few major global corporations, in collusion with our government, to poison us along with the bugs, the fungi, the weeds, and the increasingly common crop diseases. While these products are advertised as ways to control nature and our environment, we are in fact more out of control and vulnerable than ever. In our efforts to exterminate weeds, bacteria, fungi, insects, and diseases we may also be exterminating ourselves.

I am not accusing these companies of *willfully* exterminating the entire human race. That would be genocide. Although, as you will see in Part 2 of this book, many chemical companies are no strangers to the concept of genocide.

Much as tobacco companies suppressed information and test results that showed how deadly their products were, much as they lobbied and advertised intensively to defend their right to sell products that would kill people, the chemical companies have worked hard to quash research that suggests their products are harmful. And part of their defense has been a major advertising and lobbying push that insists we simply cannot survive without their products, without GMOs, without chemicals. We can't feed the world without synthetic-chemical farming, they tell us.

The bad news is that China, the land that has brought us tainted toys and melamine in baby formula, is just getting on the GMO bandwagon.[40]

From the beginning of time, we've had a love/hate relationship with nature. On one hand, nature is the source of all abundance and the resources that allow us to live and thrive— all our food, our shelter, and our enjoyment. On the other hand, nature can be brutal and cruel—earthquakes, floods, droughts, and fires have often come without warning or reason. Without protection, we are vulnerable and fearful. As a species, we remember starvation, famine, and plagues and don't want those things to happen again.

What is the antidote to fear? Control. Our use of chemicals to help us grow food stems from the primal desire to control nature and plays into our fear that we won't have enough or be safe until we control nature.

Yet, it's very obvious that we will never be able to control nature. We may be able to understand it and work with it, but the universe is much larger and more powerful than any one of us or even all of us can be. The sad truth is that we are just a small blip on the surface of time, a ripple on the ocean of life. None of which has stopped us from trying to control nature, or from trying to make ourselves immortal. (And it turns out that trying to control nature can be a very lucrative business.)

We need to begin looking at nature and this land and the soil as our allies rather than our enemies. We need to be courageous enough to change the economic and political models that reward destruction. Each and every one of us needs to be willing to take action.

The results of this great chemical experiment are in and the findings are clear, yet they have been withheld, intentionally buried or, even worse, simply ignored.

Eliminating chemicals one at a time will never make a big enough difference. The chemical industry is adept at changing names and formulas just enough to placate government regulators. By seemingly eliminating known toxins like DDT, Alar (daminozide), and lead, we are lulled into an illusion of safety. But it *is* an illusion. There is only one way to ensure our safety:

We must stop poisoning ourselves now. We must remove chemicals from the process of growing, harvesting, and preserving food.

I believe it's possible to make this essential change. When you see what it's like to be a chemical farmer these days, and just how and why we ended up in this horrible situation, you will understand that it's our moral and ethical duty to change. And the possibility for success is backed by science, government studies, and frankly, common sense.

We must demand organic; and we must do it now.

PART 2

THE TORTUOUS JOURNEY
TO OUR EXTINCTION

Thus the King conquered Nature herself.

—Flavius Josephus (AD 37–100),
referring to King Herod's Palace, which is now a ruin

3. CHEMICAL FARMING TODAY

The essential purpose of food, which is to nourish people,
has been subordinated to the economic aims of a handful of
multinational corporations that monopolize all aspects of food
production, from seeds to major distribution chains, and they
have been the prime beneficiaries of the world food crisis.
A look at the figures for 2007, when the world food crisis began,
shows that corporations such as Monsanto and Cargill, which
control the cereals market, saw their profits increase by 45
and 60 per cent, respectively; the leading chemical fertilizer
companies such as Mosaic Corporation, a subsidiary of Cargill,
doubled their profits in a single year.

—Miguel d'Escoto Brockmann, president of the 63rd United Nations General
Assembly, speaking to world leaders on the Millennium Development Goals, in
October 2008

Being a farmer is hard these days. Commodity prices soar and
fall along with the prices of gas and fertilizer and consumer
confidence, and the ever increasing amount of land needed to
earn a living makes it hard to enjoy the job. It's always been
hard to be a farmer. But a generation or two back, at least it was
a family affair that brought with it a set of values and joys that
made up for the long hours and backbreaking work.

As part of my research for this book, I met with groups of synthetic-chemical farmers around the country. These focus groups enabled me to hear, as free of bias as possible, what chemical farmers think and feel. I chose focus groups, a classic form of market research, because I felt it was essential for me to listen, to reach beyond our comfort zones and try to understand what all of us face and how we face it differently. I wasn't trying to change their minds. I was just trying to understand. And while I know lots of organic farmers, I didn't know enough chemical ones who would speak openly to me. And so I listened to farmers in Lancaster, Pennsylvania; Newton, Iowa; and Georgetown, Kentucky. They didn't know who I was or why I was there to hear them talk about their challenges and issues. To them, I was just the lady in the room taking notes (since the farmers we interviewed don't live near places that have the one-way mirrors usually used for focus groups). I was surprised to discover that farmers are highly sought after for focus groups; you have to pay them more than the usual $50 to show up. (We paid each farmer $100 in cash for 2 hours of their time, and at the end I revealed who I was and what my research was for.)

The first thing I realized is that I love farmers—even the chemical ones. There is something about their ruddy faces, rough hands, and unpretentious, stubborn natures that makes me happy. Chemical farmers are hardworking men (yes, they are mostly men) who believe they are being good stewards of the land and are trying hard to eke out a living in a very complicated system. They talk about the smell of good hay the way people in Napa Valley talk about the bouquets of fine wines (although in far fewer words). They pay very close attention to nature and respect its complexities. Chemicals do make their job easier. And the chemical companies have infiltrated their lives.

Many of the farmers I listened to rely on "crop consultants"— some of whom are independent and others who are paid by chemical companies—to help them decide what and when to spray and how to best manage their challenges. Some farmers I met with had very little education (one Amish man had received only an eighth-grade education in a one-room schoolhouse). Many farmers spoke of their "chemical dealer" as their most trusted source of information. In Iowa, most farmers belong to "co-ops" that pool their resources so they can buy chemicals and seeds more cheaply. Those co-ops are funded by the chemical companies and staffed with trained chemical agronomists who advise the farmers on what to spray and when.

The insights I gained from listening to the farmers are reflected throughout this book. But one thing is totally clear to me: Farmers are caught on a treadmill they can't quite see and don't know how to get off of. They believe they need more land to produce more, which will allow them to make more money. Chemicals enable that growth. But as you will learn, most do not grasp the classic economic principle that **the more they grow collectively, the less they will earn individually.**

The farmers constantly repeated phrases straight from the

chemical companies' brochures: We need to feed the world. We need to increase production so we can make more money off the land we already farm. If we could only export all this extra corn, our problems would be solved. (Surprisingly, they would rather see subsidies go away and they view the government and all its rules for farmers as major annoyances.)

Some are curious about organic methods, others are resentful. The number one factor that turns a resentful farmer into a curious farmer is seeing another farmer making the switch to organic and succeeding at it.

All of the farmers think growing organically is a lot harder than using chemicals. They believe they would have to reduce the size of their farms and need a lot more human help, which is very hard to get. Farms used to be worked by large extended families, and the children pitched in, too. These days, farm families are smaller, but the farms are bigger so farmers must look for outside help. They trust chemicals more than people— chemicals usually do exactly what the labels say and don't require constant supervision. At their core, farmers are fiercely independent.

One college-educated farmer who has a day job as an engineer said that he saw organic farming as the ideal, but he didn't know how to achieve it. He tried it, but then fell back into using chemicals for its ease. An older gentleman described organic as "a beautiful plan," but said he just didn't have the life left in him to figure it out. He was, however, very troubled by the dead zones in the Gulf of Mexico.

Even though farmers in different regions may grow different things, I heard many similar themes from all of the groups. (A common one: People want to move into houses near farms until they realize that farms stink.) These farmers understand what most people don't—that the reality of farming today is a far cry from the romantic view of farming from the past.

What's it like to be a chemical farmer these days? Let's take a look.

A YEAR IN THE LIFE OF A CHEMICAL FARMER

Chemical farms are in production on about 930 million acres in the United States and 3.8 billion acres globally—the vast majority of all agricultural land in the world. Currently, the synthetic-chemical system includes both genetically modified organism (GMO) and non-GMO crops, since GMO seeds have yet to be developed for many fruits, grains, and vegetables. But the chemical system and its effects are similar for both types of seeds.

Most food crops start with a seed. But the seeds today are not like the seeds farmers have used for thousands of years. Instead, farmers are strongly encouraged to choose those that have been genetically modified with the help of a bacterium— perhaps *Escherichia coli* or salmonella[1]—in order to resist the herbicide they use to keep their fields weed-free. (Isn't it interesting that those are two of the bacteria that worked?) These bacteria act as a kind of barrier for the DNA being transferred and they create antibiotics in the process (another

contributor to our overexposure to antibiotics). Billions of
dollars were spent to develop this seed, yet the government
required absolutely no health and safety testing before the
seeds were planted.

When farmers purchase GMO seed, they sign contracts
prohibiting them from saving seeds produced by this year's
crop to plant next year. The seeds are protected by patents.
This is a kind of rental agreement (and the landlord is a
slumlord), and the farmers must renew their leases each
year by buying new seeds (at higher prices). GMO-seed companies charge more for their seeds than standard hybrid[2] ones.
They are referred to as "improved" or "better" seeds by
farmers and, even more enthusiastically, by investors. By
choosing this expense, farmers commit to paying more for
seeds each year.

With this first choice of seed, farmers are choosing their
system—chemical or organic. There is no reason to choose a
GMO seed and grow it organically. The seed is designed to
thrive despite being sprayed with herbicides. GMO cotton could
be grown organically, since an organic pesticide was inserted
into the DNA of the plant, but in general, if a farmer is buying
GMOs, he has bought into chemicals. The primary GMO seeds
are corn, soybeans, and cotton, the cultivation of which
accounts for 15 percent of all farmland in the United States, but
about 33 percent of the tillable acres (farmland includes
orchards and pastures, which aren't tillable "fields").[3]

Farmers plant the seed in soil that is likely weak and
degraded from overtreatment with chemicals and undertreat-

ment with natural, organic materials. Soil, if not renewed with organic matter and treated with care, inevitably loses fertility over time. It's the equivalent of a person only taking vitamin pills instead of eating real food.

Chemical farmers believe that to make more money, they need more land. Few large-scale farmers own all the land they farm. Rather, they lease land to increase their revenues and cover the costs of their investments in machinery. So each year they look for more land to lease. Landowners who lease to farmers often don't care what the farmers do to the land so long as they can keep earning more money per acre. Absentee landowners are the worst—they expect the farmer to subsidize their urban "Lexus lifestyles."[4] Farmers do the math of how much they might earn per acre, per bushel, and per pound and gamble that by the time harvest comes around (if the weather has cooperated), they have made the right bets.

Once a farmer commits to the chemical system, he needs to buy more synthetic fertilizer to help crops grow in the depleted soil. These chemicals require lots of energy from petroleum and other sources to make, package, transport, and distribute on the farm fields. The chemicals are typically made and sold by the same companies who sold him the seeds. The chemical and seed companies are making healthy profits while the farmer struggles to pay and often falls into debt to keep this cycle going. And each time the price of fuel goes up, the prices of chemicals soar, because agrichemicals are petroleum by-products.

All crops require water to grow. In farming regions such as

California, where many of our fresh fruits and vegetables are grown, water is scarce and the soil is naturally dry. When the soil is irrigated regularly and treated with chemicals instead of organic materials, over time the salinity increases. Eventually, these soils become so salty that nothing grows in them at all.[5] It's one of the ways deserts are created.

In other areas, such as the Midwest—our "breadbasket"— there is more frequent rainfall, so farmers have less need to irrigate. But with rain comes erosion and more weeds. That water running over and through weakened soil takes the soil and chemicals with it and flows right into wells, streams, and rivers and eventually down to places like the Gulf of Mexico, the Chesapeake Bay, and all of our oceans. There, the excess nutrients meant to feed our plants cause algae to grow, starving aquatic life of oxygen and making sea life more scarce. Who tries to clean this up? The federal government, using our tax dollars.

Because farmers face serious financial risks from threats like insects, plant diseases, and weather, farmers need insurance. Federally subsidized crop insurance covers some of their risks. But many also purchase insurance through the American Farm Bureau, which claims to represent their concerns in Washington. And once again our tax dollars go to cleaning up the chemical cocktails unleashed in the environment and bailing out farmers when things go bad—which these days happens more often than not.

After farmers have bought seed, planted it, fertilized it, watered it, and insured it, it starts to grow. But so do weeds.

This is where GMO seeds really "pay off." The farmers don't need to bend over and pull weeds, or even hire people to pull weeds. Instead they spray their fields with Roundup or another herbicide. Roundup typically kills all plants—except those that have been genetically modified to resist it. For a while, the farmers are happy because their fields look nice and neat, with no weeds to compete with the crops for food and moisture. Eventually, though, the weeds start to resist Roundup, so the farmers need to spray even more. They see that weeds are becoming resistant to the chemicals they use, so they employ "chemical rotations" to try killing them by alternating atrazine and other, stronger herbicides with the Roundup.[6] Once again, the chemical companies profit the most from this stage of the process, while the farmers and the government pay the most. Before the introduction of GMO seeds, farmers tilled the land with tractors to control weeds. But frequent tilling to control weeds can lead to soil erosion. Farmers who want to qualify for federal funds under the farm bill conservation program must use techniques other than tilling to prevent soil erosion. So in a ridiculous turnabout, they use more herbicides than before, not less, thanks primarily to government regulations.

Now there's a new problem brewing. Since all the "weeds" are gone, there is less for the plant-eating birds to eat and no hospitality extended to all the good insects and birds that keep the naughty insects under control. Suddenly, the farmers have bug problems, so they need to apply insecticides to kill the bad bugs. And the chemical companies rake in the dough.

Outbreaks of *E. coli* and other foodborne illnesses have

sparked concerns and demands for new safety measures. Many *E. coli* outbreaks are linked to unsanitary slaughtering and harvesting practices, which are rampant in industrial agriculture. Some studies have shown that the unnatural diet of corn and soy fed to cattle in concentrated animal feeding operations (CAFOs) increases the *E. coli* in their digestive tracts.[7] The chemical industry—in an effort to escape blame for its own foul deeds—has convinced the government that wildlife is the cause of disease outbreaks and are calling for "sterile" farms![8] For centuries, farmers have relied on hedgerows and wetlands to keep their land healthy by preventing wind erosion and providing habitat for beneficial animals and insects. Those are now being forceably removed to prevent wildlife from walking through the crops, since most wildlife make their homes in the hedgerows and wild areas on and near farms.

The weather is a risk factor, too. Add a flood or drought— you pick—and the problems multiply. A flood causes more run-off and erosion, and the wimpy root systems typical of chemically fed crops can't maintain their tenuous grasp on the soil. In a drought, irrigation can prevent some crop loss, but not all. Chemically farmed plants are more fragile when they don't have enough water because their weak root systems aren't strong enough to find water. Either way, the farmer must rely on insurance to cover his losses, banks to cover his costs, and the government to save his ass.

Insurance premiums are just one factor in the massive debt farmers accumulate. The giant tractors they use (and that slow traffic for the rest of us) cost hundreds of thousands of dollars. Few chemical farmers ever pay off their debts.

Now it's time to talk about sex. All creatures are pro-grammed from birth to reproduce, and reproduction happens with sex, even in plants. Without sex, there are no corncobs, no soybeans, no fruits, no vegetables, no food. GMOs are like one giant pandemic of sexually transmitted disease. Pollen from these plants drift on the wind into non-GMO fields and do what comes naturally—procreate. A perfectly good organic field becomes contaminated—raped, if you will. Chemical companies—Monsanto is renowned for this—have claimed that farmers who saved seeds from GMO crops stole their "intellectual property" and sued them for damages, in some cases even when the farmer swore he never planted them in the first place.[9] In recent years, Monsanto has filed at least 100 lawsuits around the country related to this "theft."[10]

You can't see the difference between a GMO seed and an organic one that's been contaminated. The farmers being sued lose—not just financially, but also because they must scrap seeds they may have cultivated and bred over many years to produce the best yield for that farm. (It reminds me of how rape was often considered the victim's fault—and she was the one who suffered the shame and punishment for it.)

Fortunately, "terminator seed" technology is prohibited. Terminator seeds were GMOs that became infertile after 1 year. Or made celibate, if you will. But sex is essential to our survival. "If not for sex, much of what is flamboyant and beautiful in nature would not exist. Plants would not bloom.

Birds would not sing. Deer would not sprout antlers," writes Olivia Judson, PhD, an evolutionary biologist, in her book *Dr. Tatiana's Sex Advice to All Creation.* "For reasons that remain mysterious," she adds, "the loss of sex is almost always followed by swift extinction. Apparently, without sex you are doomed."[11]

Finally, harvest time comes. Farmers growing crops that can't be picked by machine (such as some fruits and vegetables) have to hire help, many of them migrant workers. They are often illegal immigrants who have no health insurance, and they bear the brunt of the damage caused by toxic farming practices, with much higher rates of cancer, birth defects, and other diseases. But because they are just itinerant laborers, their problems are not the farmers' responsibility. I have heard about farmers actually enslaving illegal immigrants to do their dirty harvesting work for free.

Farmers that don't need laborers sit alone in the cabs of their air-conditioned tractors, driving back and forth for hours and maybe days, listening to the radio, eating sandwiches, drinking coffee. They may remember hearing stories when they were growing up about how harvesting used to be a joyous time when neighbors came together to help each other and then celebrate. Now, to tell the truth, farmers may be kind of lonely. Their giant, debt-inducing machinery harvests the crop and takes it to market. Since most crops are a commodity, they sell for the going rate, whatever that may be. No matter how much time and sweat farmers put into growing their crops, they get what the market says it is worth.

Though the whole biofuel craze has raised the demand for corn, for instance, at the end of the day farmers still are falling behind in their payments. Farm subsidies may help a bit, but the complicated effort it takes to figure it all out depresses them, they say, since it's never enough to solve their problems for good.

A person who works a lot with farmers told me there is "no graceful way out" of this chemical system. Chemical farmers who consider switching to an organic method will encounter many disincentives. Government subsidies will be withdrawn, and they may owe too much money to too many people to make the risk worth the effort. Bankers would scoff at the idea and maybe even forbid them to transition to organic. Plus, it takes 3 years of "getting off chemicals" and cleaning up in order to become certified organic. During this time, yields decline as the soil recovers from years of abuse and the farmer struggles to learn a whole new farming system with only negligible government assistance, insurance, and other help to get through those 3 years. This very difficulty is why we need to make substantial changes quickly.

What about the food itself? Only a tiny percentage of GMO crops go directly from the fields to our tables. Most of the corn and soybeans grown in the United States are made into sweeteners (particularly high-fructose corn syrup), animal feed, cereals, crackers, bread, chips, and fuel. Increasingly, these crops are being exported to other countries, putting millions of farmers in other countries that don't share our overly generous subsidy programs out of commission and into starvation.

Corn and increasingly soy are used to feed livestock, which are kept in large feedlots to fatten them up. Eventually you consume the corn when you order your giant steak at the local steakhouse or a cheap burger at a fast-food restaurant. If any by-products of our corn-based farming are left over, they're sent to public school systems to feed our children.

What happens to all the waste, the leftover plant and animal material from our farms? Little that is natural is returned to the soil to be composted. Most farm waste (including dead animals) is fed to animals or sold for fuel. Some farmers fertilize their fields with sewage sludge, or municipal waste. It contains human waste, which is contaminated by pharmaceutical drugs and cleaning supplies that have literally been flushed down the toilet, as well as industrial waste.

In concentrated animal feeding operations (CAFOs)—these are more like food factories than farms—thousands of hogs, cattle, and poultry are fattened before slaughter. Their waste is funneled into large slurry ponds, and it tends to run off into water supplies and emit methane, which is another potent greenhouse gas and a major contributor to our climate crisis.

Rather than use the waste, as nature does, to reinvigorate the soil after a hard growing season, the fields are left bare. They get neither a warm layer of compost nor a protective cover crop. Winter winds and rains lead to more erosion. Any mycorrhizal fungi that might have been in the soil storing carbon have been decimated by fungicides, and the few that remain starve without cover-crop roots to live on during the winter. The landscape is desolate. There are no seeds to save. In fact, the "gene police" from Monsanto may slip into fields throughout the season to make sure that all the rules are being followed—and they won't hesitate to sue if they believe there has been an infringement.[12] Winter is long and dark for farmers, who know that they have to start the cycle all over again the following spring.

Throughout this synthetic-chemical cycle, fertilizers, herbicides, and insecticides run off the fields, as does topsoil, polluting freshwater sources and damaging the health of the country and the environment. The chemical system puts much more carbon into the air than it takes out. Meanwhile, no one has tabulated the long- or short-term health consequences of farming with chemicals—not just in the United States, but around the world. Farmers are just barely surviving, the government and the American people are left to pay the bill for cleaning up the environment and health care costs that are out of control, and our relationships and reputation as good global citizens are seriously jeopardized. The only parties who are walking away from this scenario better off financially than they were when they began are the global GMO seed and chemical companies, and the lobbyists who feed off of them.

It all sounds like a postapocalyptic science fiction story, yet all of this is happening right now. We just haven't connected the dots. We are bombarded every day by bits and pieces that don't seem to add up and leave us feeling like we can't understand or control the problems, let alone solve them.

FARMERS ARE ADDICTED

What do farmers in America think of Monsanto, Syngenta, and the other agrichemical companies? I would describe it as a codependency with a chemical addiction as the glue that binds them together. They think a lot less of Monsanto since they got them hooked on Roundup Ready GMO seeds and then "jacked up the price of seed."[13]

Few of the farmers caught in this deadly cycle believe they are doing anything wrong. They love their big tractors and take pride in their beautifully neat and weed-free rows. They view their work as feeding the world and sustaining the legacy of the Great American Farmer. But beneath these positive feelings, many farmers confess that they long for the time when farming was more manageable, more profitable, and more family oriented.

Dan Wiese, the facilitator of my farmer focus groups, has been talking to farmers and their wives for more than 20 years. He's been hired by almost every chemical company to find out what farmers need and how to sell it to them. He's from Iowa, and he insists that GMOs save money and time. But what they've done to the social system of Iowa, he believes, is a crime. Chemicals and GMOs have both enabled and forced farms to grow ever larger, causing population densities to fall so low that small towns wither and schools must consolidate.

"When the school is gone the town begins to die, too," Wiese says. "It's like taking the heart out." He also lamented the Iowa flood of 2008. The flood before it crested at 19 feet. In

2008, the flood's high point was 32 feet. About 9 square miles in eastern Iowa was underwater, and much of Cedar Rapids is still uninhabitable. Yet we fail to make the connection between how chemicals destroy the soil's ability to absorb and hold water and the increase in devastating floods.

In my research on farmers, I was especially interested in visiting Kentucky.

It has fewer certified organic farmers than any other state, according to the USDA, and it's one of the few states where the number of organic farmers is *declining*, dropping from 23 in 2006 to just 3 in 2007.[14] (Fortunately, in most other states the number of organic farmers is increasing.)

In a Comfort Suites conference room right off the highway in Georgetown, Kentucky, under a glaring fluorescent light, eight farmers, Wiese, and I talked about farming. One woman attended, although she stated right up front that the only reason she farmed was that she loved her husband and he loved to farm.

After the discussion, I concluded that Kentucky agriculture is in a state of transition. Its main cash crop has traditionally been tobacco, and there isn't much demand for organic tobacco, though some is exported to Germany. That must be grown by the three certified organic farmers "up in one corner of the state," ventured one farmer in the focus group. Many farmers in Kentucky still grow tobacco because no other crop generates as much money per acre, even though it is an extremely chemical- and labor-intensive crop. Tobacco companies are very involved in (some might say they dictate) how farmers grow their tobacco and which chemicals must be used. To get a contract with a tobacco company, farmers must agree to comply with their exacting demands.

Many Kentucky farmers are switching to vegetables and hay to feed the still-growing (although struggling) horse racing industry, an important market in Kentucky, the home of Churchill Downs. They joke that horse owners want hay and alfalfa that are green in color, not realizing that the weather in which it is grown and dried, rather than the quality, dictates the color. Many stables pay a fortune to have the green stuff shipped in from out west, "because their horses won't eat it if it isn't really green," despite the fact that local hay is only a couple bucks a bale. The farmers all had a good laugh about how the recession had suddenly "changed the horses' appetites for local hay."

These farmers, like all farmers, struggle most with the dramatically falling prices for their crops, while the costs of producing them are going up. And there seems to be no end in sight.

Many of the Kentucky farmers have been working the same land for generations. One elderly gentleman, who could trace his farm back to the 1700s, expressed concern about contaminated wells. He sits on the water board for his town, and the levels of some chemicals are so high that residents have been urged to drink bottled water. His perspective was that "this is just like the fall of Rome, where the Romans had exported all their farming and spent all their time obsessing about sports."

When I revealed who I was and what I was doing, the farmers were surprised and curious. One young man had attended culinary school but couldn't find a job, so he was working 60 hours a week managing a local deli and growing vegetables on the side. After he found out what I was working on, he pulled me aside and said that the main reason he used chemicals was as a form of insurance. There are so many risks in farming, he said—weather, bugs, prices, and diseases—and using chemicals made him feel less vulnerable to those risks.

Another farmer asked me why, if organic is so much better for you, he doesn't see any advertising saying that. I explained how the USDA's rules for organic labeling prohibit companies from making that claim. "If that yogurt company [whose products] supposedly make your digestion better can make that claim, why can't an organic farmer?" he asked. That's a good question!

After I returned home, I was told by a friend that Kentucky's biggest cash crop is now marijuana. Kentucky is the second biggest producer of pot in America, right after California.[15] You certainly wouldn't want organic inspectors coming on your land if your primary source of income was a certain illegal weed.

THE TRUE COST OF DENIAL

What about farmers in the rest of the world? Again, the Web sites of chemical companies would have you believe that without GMOs and chemical crop protection, there will be mass starvation and environmental destruction. In fact, the exact opposite is true.

Our government (with its arguably well-meaning subsidies to farmers), combined with the chemical companies, are the two bullies in the alley who have delivered to farmers everywhere the one-two punch.

Let's take a look at India. For years the price of cotton was low around the world because there was more cotton than people needed—and most companies buying cotton choose the lowest-priced source. Then US farm subsidies artificially reduced the price of American cotton and suddenly, the cheapest cotton was coming from America, not India. Even though the cost of living in India is a fraction of what it is in the United States, suddenly Indian farmers couldn't make a living. That was the first blow.

Now you have the follow-up blow. Desperate Indian farmers get calls from companies using Bollywood movie stars and Hindu deities to help sell the farmers on "magic" seeds. With literally nothing to lose, Indian farmers borrow money to buy the seeds and the companion chemicals, never realizing that the price will increase each year. After the first year, they find out that it costs much more to maintain their crops due to the ever-increasing prices of seeds and chemicals. Yet they are still plagued by insects and, like all promises of magic, the yields are disappointing at best. Before long, the money lenders are knocking on their doors and there is not enough revenue from the crops to pay the debts.

More than 160,000 Indian cotton farmers have killed

themselves in the past decade. The favored method of suicide? Ingesting chemical pesticides.[16, 17]

The International Assessment of Agricultural Knowledge, Science and Technology for Development, with the urging of and $12 million in funding from the World Bank and the United Nation's Food and Agriculture Organization, was an unprecedented survey of agriculture around to world designed to determine the best solutions for feeding the world. More than 400 scientists and 30 countries participated in the proceedings, including nongovermental organizations and nonprofits from those countries who have on-the-ground experience working with farmers. The resulting reports, released in 2008, looked at many aspects of farming, from reducing hunger to improving rural livelihoods to long-term development that is healthy for people and the environment. It also looked at the impact of government subsidies that encourage farmers to dump the surpluses from countries like the United States on the global market and destroy local markets. The survey report's sweeping and comprehensive recommendations include returning to more "natural" and traditional farming methods (away from GMOs and chemicals). Giving women the right to own the land they farm was another radical recommendation (women still aren't allowed to own property in many countries). Building roads so farmers can get their crops to market is a simple but obvious idea. Finally, they recommend ending subsidies.[18]

Not surprisingly, a representative from Syngenta, Deborah Keith, PhD, Syngenta's crop protection research portfolio manager, walked out of the meeting on the recommendations before it was completed. In a commentary published in the April 2008 issue of *New Scientist,* she explained why.[19] The facts on organic farming she cited were straight from the Syngenta Web site, which at the time claimed that it would take three times more land to feed the world with organic farming practices, and that organic farming offers no health benefits. (Syngenta has since changed the information offered on its Web site.)

"Sadly, social science seems to have taken the place of scientific analysis," says Keith about the recommendations from the international survey. "Social science" is often invoked when an idea is being discredited. To many people, social science is unfairly perceived as one step removed from pseudoscience and a step and half from quackery. The complexity of human behavior and the value of a human life have no place in business. And another revealing comment from Keith: "Innovation is only created through investment, and investment must be rewarded— another seemingly obvious fact which was overlooked."[20]

In the Akola region of Maharashtra, India, where there were 5,000 suicides from 2005 to 2007, a local textile company started contracting with a few hundred small farmers to grow organic cotton for them. The textile company pays a fair price and trains the farmers how to grow organically. The farmers seem happy and the textile company has been able to provide organic cotton fabrics to meet the growing global demand.[21]

There have been no farmer suicides since the program started.

That is social science.

4. THE BIRTH OF OUR CHEMICAL ADDICTION

It is a common error in Europe, to look on all natives not reduced to a state of subjection, as wanderers and hunters. Agriculture was practised on the American continent long before the arrival of Europeans. It is still practised between the Orinoco and the river Amazon, in lands cleared amidst the forests, places to which the missionaries have never penetrated. It would be to imbibe false ideas respecting the actual condition of the nations of South America, to consider as synonymous the denominations of 'Christian,' 'reduced' [meaning living at a mission], and 'civilized;' and those of 'pagan,' 'savage,' and 'independent' [meaning not living at a mission]. The reduced Indian is often as little of a Christian as the independent Indian is of an idolater. Both, alike occupied by the wants of the moment, betray a marked indifference for religious sentiments, and a secret tendency to the worship of nature and her powers. This worship belongs to the earliest infancy of nations; it excludes idols, and recognizes no other sacred places than grottoes, valleys, and woods.[1]

—Alexander von Humboldt (1769–1859)

How *did* we get here? How did we get to this place that leaves our lives and our children's futures at such grave risk? How did we create this world where man-made chemicals have invaded every aspect of our lives and threaten our climate, our health, and our futures? How did we come to believe that we cannot survive without the help of artificial, toxic, and harmful products on and in our food? Are we destined to become an abandoned planet by committing "ecocide"?[2]

In part the answer to this question lies in our passionate but damaging love affair with chemicals.

Most people think that our dependence on synthetic chemicals (which I define as toxic substances that are *manufactured* for a specific purpose) to grow food began in the 1940s or 1950s, or perhaps in 1842 with John Bennet Lawes's discovery of the first synthetic superphosphate fertilizers. But our dangerous liaison with chemicals started even earlier than that.

The Chinese were using arsenic sulfides in agriculture as early as the year 900.[3] By the 1300s, arsenic, lead, and mercury were being used in Europe for all sorts of things, including medicines.[4]

The work of a German chemist named Justus von Liebig (1803–1873) created the great divide between "book" chemical farming and experiential "biological" farming. In the early 1800s, chemistry was still viewed as a "pseudoscience."

"Chemists and alchemists were still the laughing stock of scientific societies throughout the 1700s," writes author Will Allen in his book *The War on Bugs.* "Scientists and other intellectuals frequently referred to them as quacks."[5]

Liebig believed that man and science could replicate the

most valuable resources of nature—sugar, salt, and nutrients—
so that man would never again feel vulnerable to the whims of
nature. He asserted that the stuff soil is made of—humus or
organic matter—was not necessary as long as we could replicate its mineral and nutrient content and that man-made
minerals and nutrients would be superior to nature's[6] because
after all, they were made by man, who was superior to nature
in every way. The study of chemistry had started as curiosity
about the nature of the stuff of life and then developed mechan-
ical arrogance—the "life is just a machine" paradigm that we
still often live by today.

"It is the greatest possible mistake to suppose that the
temporary diminution of fertility in a soil is owning to the loss
of humus; it is the mere consequence of the exhaustion of the
alkalies," Liebig proclaimed in his seminal book *Organic
Chemistry in Its Applications to Agriculture and Physiology*, first
published in 1840.[7]

He was the originator of the idea (although not the slogan)
"better living through chemistry," and he initiated the trend of
trusting academic scientific research on agriculture over the
experience of farmers. His laboratory and "scientific" methods
inspired many Ivy League and land-grant university agricultural programs and are still used today, even though many of
his original theories were long ago debunked.[8]

A German, Liebig accused England of "robbing all other
countries of their fertility" by stealing the bones of the dead
from the battlefields of Waterloo in Belgium and the Crimea
in a craze for fertilizer: "Like a vampire she hangs from the

neck of Europe."[9] And it's true: England did harvest bones from whatever sources it could. Ground-up and buried bones had long been considered important sources of soil nutrients.

In the 1800s, the idea of staying on one's land was highly appealing. Great Britain was in its heyday as an empire. Americans had established vast plantations and estates. The ruling classes had invested lots of money in big homes and big cities, and the desire to control nature by improving the production of natural resources was strong. Better yet, creating essential resources through chemistry and making money from it seemed like a great idea. The stigma among the upper classes against engaging in trade was falling away, and educated young bucks were looking for ways to make their fortunes.

Liebig and his set truly adhered to the evolutionary vision promoted by the anthropologist Lewis Henry Morgan. Morgan believed that people could be categorized into three evolutionary groups: the savage, who subsist on hunting and gathering; the barbaric, who use primitive agriculture and domesticate animals; and finally the civilized, who have the ability to write and are at the peak of the social hierarchy.[10] Liebig and friends placed themselves at the top of the civilized world.

These men believed that it was their God-given right to control nature, and they had no respect or appreciation for anything beneath them in the evolutionary hierarchy, whether it was the lower classes, other races, women, or nature.

A TROJAN HORSE FILLED WITH FECES

Ironically, it was a totally natural and "organic" material that began our reliance on external agricultural inputs and paved the way for our chemical addiction.

In 1804, the German botanist and explorer Alexander von Humboldt brought back a sample of guano from his ambitious adventures in South America. Off the coast of what is now Peru, von Humbolt encountered islands on which thousands of years of seabird feces had made giant cliffs of concentrated and powerful nitrogen-rich powder, or guano. The indigenous people had used it for generations to fertilize their crops—as a traditional proverb says, "Guano, though no saint, works many miracles."[11]

The Incas ranked *huanu* "right alongside gold as gifts of the gods and forbade anyone on penalty of death to molest nesting birds."[12]

The sample von Humboldt brought back from his travels was analyzed by a few different chemists and found to be high in nitrogen, an element prized by farmers for its ability to improve the fertility of their soils.

Until then, farmers had relied on recycled urban and farm waste (manure from livestock and people, wood ash, and bones—including bones from the battlefield at Waterloo) to fertilize their fields. In 1840, W. J. Myers and Company imported 20 casks of guano for experiments. After a few farmers in Great Britain achieved near-astounding results from applying guano, demand grew quickly.[13]

In 1850, 80 percent of Americans lived on farms, which
accounted for 75 percent of the nation's economic output.
Farmers who had gotten fat and lazy off the fertile virgin lands
(often on the backs of slave labor) started seeing the inevitable
decline of fertility that went along with their "primitive agricultural techniques,"[14] such as growing the same crops on the
same land for more than a hundred years without rotation and
without adding back sufficient organic material. Competition
from other farmers who were farming fresh land made East
Coast farmers even more determined to find ways to improve
their yields. Guano fit the bill and, influenced by strategic
advertising and editorials in farm magazines of the time (many
of which were affiliated with farm stores that benefited greatly
from guano sales), farmers started buying the fertilizer.[15]
Suddenly, in addition to the smell of bird droppings, people
started to smell profits, and the "Great Guano Rush" was on.

Thus begins one of the more unusual forgotten episodes of
our past. By 1851, 66,000 tons of guano had been imported to
the United States.[16] Prices fluctuated wildly, but mostly rose
higher and higher, and opportunists began marketing inferior
guano from other islands and making fortunes. (One of the
companies that first made a fortune from guano was W. R.
Grace and Company, which much later went on to make profit
from manufacturing asbestos.) In 1850, President Millard
Fillmore, in his first State of the Union address, stated that
"Peruvian guano has become so desirable an article to the
agricultural interests in the United States that it is the duty of
the Government to employ all the means properly in its power

for the purpose of causing that article to be imported into the country at a reasonable price."[17]

In 1856, the US government passed the Guano Island Act, which stated:

> Whenever any citizen of the United States discovers a deposit of guano on any island, rock, or key, not within the lawful jurisdiction of any other Government, and not occupied by the citizens of any other Government, and takes peaceable possession thereof, and occupies the same, such island, rock, or key may, at the discretion of the President, be considered as appertaining to the United States.[18]

As a result of that act, the US government seized 94 islands off the coast of Peru just to harvest bird shit.[19]

Few Americans wanted to harvest guano, so Chinese slaves were imported and Peruvian Indians were imprisoned to get them to do the work. Many workers were kidnapped into slavery and forced to work "twenty or more hours a day, six days a week, to fill quotas of four to five tons each, for which they were paid three reales (about a third of a peso), two of which were withheld for meals," according to one historical account.[20] The only escape was death, and many workers, if they weren't buried alive in collapsed guano trenches, threw themselves into the sea.

Despite the tragedy of how guano was acquired, it did replenish the soil depleted by a century of nonstop cultivation.

Better farming practices, such as crop rotation, and applying compost and free sources of fertilizer such as livestock manure would have worked just as well as if not better than guano. But, as is often true today, farmers of that era believed that a solution they had to pay for must be better than the one that required their own hard work. In contrast to the slow and steady hard work of soil management, guano was a quick fix that prepared farmers for the chemicals to come.

By the 1870s, all natural sources of guano were exhausted. But the economic infrastructure and the expectation that fertilizer needed to be bought from some external source remained.

"The most important changes were in the thinking of farmers," observes Richard A. Wines in *Fertilizer in America*.

> In the areas where guano was first introduced, farmers were already purchasing urban manures and soil supplements, but guano greatly reinforced and extended this commercial mentality. Thus, farmers became accustomed to purchasing large amounts of relatively expensive, concentrated fertilizers. Farmers also learned to accept fertilizers that acted rapidly and whose strength was used up in a year or two, as contrasted to lime, ashes, bones, or stable manure, whose effects were visible for much longer periods. Most importantly, as a result of the use of guano, they came to expect that any other fertilizer they tried should be as highly concentrated.[21]

Many companies sold "guano" fertilizer for years after all the natural guano was gone (it was typically made from inferior manures from other regions). The substitutes were much cheaper, but also much less effective. Meanwhile, during the approximately 30 years that guano was in use, farmers became dependent on external resources to farm their land and forgot that they could use their own natural resources. By the time they realized this new guano didn't work, it was too late. They were hooked.

ARTIFICIAL FERTILIZERS EMERGE

As the demand for commercial fertilizers continued to grow, John Bennet Lawes was developing superphosphates—artificial, concentrated nutrients that delivered both nitrates and phosphates, both of which are critical to plant growth. In contrast to Liebig, Lawes considered himself a scientist (rather than a chemist) and was committed to improving and understanding farming as it actually occurred in nature, not in a laboratory. His test plots and scientific experiments continue to this day at Rothamsted Research in England.

Though Lawes made a fortune selling his chemical fertilizers, he continued to believe that manure was essential for farmers. "The repeated declaration of chemists that farmers will be able to grow as fine crops by the aid of a few pounds of some chemical substances as by the same number of tons of farm-yard dung, is never likely to be realized," Lawes stated in a leaflet published in 1846.[22]

Lawes sold his business in 1872[23] but continued his agricultural experiments at Rothamsted. And at the end of his life, long after he had sold his business, he took a stand against manufactured fertilizer. "I do not consider that artificial manures are very suitable for the growth of garden produce," he wrote. "If I were going to establish myself as a market gardener I should select a locality where I could obtain a large supply of yard manure at a cheap rate."[24] Even the inventor of artificial fertilizers recognized that they were inferior to good old-fashioned manure.

But by then too many new businesses were making too much money selling artificial fertilizers to farmers. And farmers were becoming increasingly desperate to find easy ways to increase their production—they were hooked on the fast results achieved with guano.

New miracle products for farmers were introduced and marketed aggressively. Until the 1880s, the fertilizer industry was rife with fraud and inconsistent and ineffective products. Most were still primarily made from naturally occurring and organic materials such as fish scraps, mined phosphate deposits, and bones. By the end of the century, fertilizer manufacturing had become an important source of revenue for the nation's largest meatpacking companies, Wines reports.[25]

Sodium nitrate, a by-product of salt mining (and also a powerful explosive), potash and sulfur, and waste from coal, coke, and oil processing were the main ingredients of the growing global fertilizer industry, which took root at a time when there were few government regulations, no testing for health problems, and no consequences for selling toxic poisons.

"As the Industrial Revolution progressed, even more enormous quantities of waste chemicals accumulated," says Will Allen in *The War on Bugs*. "Finally, after creating mountains and filling valleys with toxic waste, the industrialists were forced by state and national regulators to dispose of it or face serous fines and disposal fees. When the state and national regulators began imposing fines, most of the mining and manufacturing corporations followed the lead of the German potash syndicate and turned to agriculture as the major dumping ground."[26] Rather than bear the expense of disposing of their own toxic garbage, these companies charged farmers to do it for them, and in the process they turned our agricultural lands into the primary dumping ground for industrial waste.

This is still happening today. The EPA, in partnership with the American Coal Ash Association and the Electric Power Research Institute, has approved an effort to *increase* the use of coal ash in agriculture. Already more than 180,000 tons of coal ash, which is loaded with mercury and arsenic, are used on agricultural fields every year.[27]

A LITERAL EXPLOSION OF CHEMICALS

Germany, meanwhile, was building a chemical industry that still is the world's strongest. With Liebig as their inspiration, the Germans invested heavily in the education of chemists and were flush with minds focused on how to replicate artificially all the natural materials that they couldn't get easily from within their own borders, including and especially fertilizers.

In 1909, chemist Fritz Haber figured out how to "fix" nitrogen from the air. The air we breathe is 78 percent nitrogen. Pulling it out of the air to put it directly on crops became a national obsession for Germany. Haber's work was funded by BASF (a chemical company that still operates today). In partnership with Carl Bosch, he was able to commercialize the process of creating artificial nitrogen fertilizer (which was called the Haber-Bosch process) and start earning money from it.

But nitrogen has purposes other than just fertilizing plants. Nitric oxide also is a key ingredient in explosives; a quaint 1950s promotional brochure for the DuPont company notes, "Oddly enough, nitrogen, the element which sustains life in the human population, is just as vital for military explosives."[28]

Until the introduction of the Haber-Bosch process, explosives came from saltpeter, a naturally occurring form of potassium and sodium nitrate found mainly in Chile. During World War I, Germany's adversaries blocked it from getting saltpeter from Chile. The German government subsidized factories to build a chemical substitute using the Haber-Bosch process of producing ammonia, which was then oxidized to make the nitrates essential to synthetic fertilizers . . . and explosives.[29]

After the war, the victorious Allies briefly considered bringing Haber to trial as a war criminal because of his development and oversight of the first military use of poison gas. But he "grew a beard and hid in Switzerland for a few months until the fuss died down," says Diarmuid Jeffreys in his book *Hell's Cartel.* Haber later returned to Germany to rebuild the chemical industry and won a Nobel Prize for his discovery of synthetic ammonia in 1918.

American companies such as DuPont also made both explosives and artificial fertilizers. But the Germans still reigned supreme, thanks to their government's strategy of funding chemistry departments at universities. DuPont invested both money and time in appropriating trade secrets from the German chemical companies.[30]

Very few government regulations limited the chemical companies' damaging practices. In fact, World War I had added an element of patriotism to the development of new chemicals and the products made with them. The great industrialists were all making fortunes, and they controlled the fate of the world with their economic might. They weren't about to let a few bugs get in their way.

EXPERIMENTS IN EXTERMINATION

Though the chemical pesticide industry had existed for some time, it didn't truly come into its own until chemical fertilizers took off. Then, however, it began spreading like a plague of locusts. Arsenic, lead, and cyanide were sold as bug and rodent killers, which were often developed and sold by the same companies that marketed these ingredients as "medicines."[31] For instance, Bayer, the German company that first marketed aspirin, was integral in the development of chemical pesticides.

Note that farmers didn't just roll over and accept easily these new pesticides. Farmers were rightfully skeptical and reluctant to use them—they were quite familiar with the toxicity of many of these chemicals. But they gradually achieved widespread acceptance with clever marketing and aggressive business tactics.

The German chemical companies Bayer, BASF, Hoechst, Agfa, and Weiler-ter-Meer joined together in 1925 to form the cartel IG Farben to protect and control their companies' businesses and trade secrets. The cartel made competition exceedingly difficult and its member companies highly profitable and operated many subsidiaries both covertly and openly in the United States and other countries. The German business leaders based their cartel model on the Rockefellers' Standard Oil monopoly. (Standard Oil also sold pesticides as well as oil and actually partnered with IG Farben on a number of projects, including the production of leaded gasoline.)

Both IG Farben and the American chemical giants used their profits to advertise their products, focusing on the "war" on bugs and likening farmers to a patriotic soldiers who must protect their farms and families from evil threats, tapping into a primal social fear of not being patriotic or *manly* enough to protect one's family.[32] (Today, many agricultural chemical products still have brand names such as Warrior, Prowl, Pounce, and Bullet that reflect this marketing approach.[33])

Meanwhile, the high prices of chemicals put a lot of small farmers out of business.

The chemical companies also employed the humor and style of Theodor Geisel, later renowned as the children's book author Dr. Seuss. In the late 1920s, before he published his first book, he designed advertisements for Flit, an insecticide produced by Standard Oil. His lighthearted approach led to greater acceptance by consumers, persuading them that using chemicals to grow their food was both safe and necessary.[34]

Later, as Americans marched proudly into World War II to defend our freedom and save the world from Hitler, we were using many of the same chemicals in America that the Nazis were using to exterminate the Jews—and we had been buying the chemicals from Germany.

Zyklon B (or Cyclone B), the infamous poison gas that the Nazis used to execute millions of people, was a cyanide-based pesticide[35] used for "delousing" that was developed by Fritz Haber. The ethnically Jewish chemist had converted to Christianity in 1892 and was a German patriot, but he left Germany for England in 1933, having been isolated from the scientific community for his role in developing agents of chemical warfare. His conversion and service to his country were not enough to save him from the rising tide of anti-Semitism. Fortunately for him, Haber did not live long enough to see the extent of the destruction his inventions caused. (Haber's first wife, also a chemist, so strongly disagreed with his work on poison gases, which he tried to hide from her until an explosion killed her dearest friend, that after an argument following a dinner party she took a gun, went out into the garden, and killed herself.)

Zyklon B was manufactured and distributed by Degesch, a subsidiary of IG Farben, who supplied the necessary materials and apparatus to two licensed producers.[36] The German chemical companies that were part of IG Farben planned to build a synthetic rubber "factory" next to the concentration camp at Auschwitz, and one of its subsidiaries was to supply gas for the camp's mass executions. The brutality of the overseers at the factory was horrific, often exceeding the brutality of the SS at the concentration camps. "No mercy was shown. Thrashings, ill treatment of the worst kind, even outright killings were the norm," one of the very few surviving inmates said. "The murderous working speed was responsible for the fact that, while working, many prisoners suddenly stretched out flat, gasped for breath, and died like beasts."[37]

IG Farben also funded the infamous Dr. Josef Mengele, who was conducting "research" on many of the pharmaceutical drugs that IG Farben and especially Bayer and its own doctor, Helmuth Vetter, were developing. Vetter, the camp physician at Auschwitz, enjoyed his work (to put it mildly). "The experiments were performed. All test persons died," Vetter wrote to colleagues at Bayer headquarters [in Leverkusen then and still today]. "I have thrown myself into my work wholeheartedly. . . . I feel like I am in paradise."[38]

More than 1.5 million people were killed at Auschwitz by IG Farben poison gas and the torture of slave labor. No rubber was ever produced at the factory because the forced labor never finished building it. And only a handful of the thousands of Jews sent to work in the factory survived.

WHO REALLY WON WORLD WAR II?

After the war, many IG Farben executives were tried for war crimes. Only three were convicted of slavery and mass murder, but they were released from prison after a few months. All three went back to work leading their businesses, which experienced unprecedented growth after the war. The Allies eventually broke IG Farben into three companies: Bayer, BASF, and Hoechst, now known as Sanofi-aventis.

Today the pharmaceutical industry remains entwined with agrichemical businesses, an association perhaps best illustrated by Bayer CropScience, the world's second-largest agricultural chemical company, about which they say, "If people turn to Bayer for what ails them, why can't plants?"

BASF calls itself the world's leading chemical company and it is the largest producer of chemicals in the world today, Sanofi-aventis is a global pharmaceutical company.

After World War II, American companies acquired the German companies' trade secrets and began selling those same products to American farmers by using patriotic advertising. Companies such as DuPont, Monsanto, Dow, American Cyanamid, Eli Lilly, and various cigarette manufacturers all were involved in the highly profitable poison business. Most of these companies knew that their products were causing cancer, especially in their employees, says Devra Davis, PhD, in *The Secret History of the War on Cancer*. Yet the companies' leaders actively, intentionally, and repeatedly denied, covered up, and used doubt and ridicule to prevent closer scrutiny or government regulation of their products. They even set up academic and research institutes to disseminate information giving a favorable view of their businesses to the public. After all, a great deal of money was at stake.

Americans were not fearful or suspicious of chemicals in the aftermath of the war. Rather, they transferred their pride in our military might to chemical agriculture. Ironically, what was artificial came to seem "normal," safe, and conventional. To a large degree, this thinking remains unchanged today.

Walk into many supermarkets today and you'll see labels above the food that is *not* organic—that is, food that has been grown with chemicals, often from GMO seeds—calling it "conventional." Conventional cherries, conventional corn, conventional whatever. To most people, the word "conventional" is a nonthreatening term that implies safety and adherence to traditional standards. How has a term that means time-tested, true, and safe come to be used in association with a method of farming that is anything but? (That's why I refuse to use it when referring to chemically grown food.)

I have not been able to trace the origin of the use of "conventional" to denote chemically farmed food, although my friend George Bird, PhD, of Michigan State University, claims responsibility. He was concerned by synthetic chemical agriculture being called "traditional" agriculture, so he started substituting "conventional." What I do know for certain is that how we label things is extremely important to how they are perceived. The new buzz phrase that Syngenta, Bayer, Dow, DuPont, Cargill, Archer Daniels Midland, and Monsanto use to describe their chemicals and biotech product lines is "crop protection." Like conventional, "crop protection" sounds so safe, like you have a security force out there guarding your food from invaders. These companies pervert our language to mislead us. We fall for it all the time.

We organic food activists love to pick on Monsanto. But until we address the problem of agricultural chemicals, picking on one company alone will just be a distraction. However, Monsanto is a good example of how a company—indeed, a whole industry—morphs and changes over time to continue to reap profits from very destructive products. Like Hydra, the mythical beast that sprouts new heads to replace those that Hercules cuts off, the chemical companies come back stronger and more dangerous each time. Even more discouraging, you can see over the years how they keep trying to correct themselves by getting out of bad businesses and into new ones. Unfortunately, almost all of their businesses are just as bad, if not worse. (Where is Hercules when you need him?)

Monsanto started producing saccharin in 1901, followed by polychlorinated biphenyls (commonly referred to as PCBs) and Agent Orange (both are primary producers of dioxins—which are extremely carcinogenic and also powerful hormone and endocrine disrupters). Then the company developed Roundup in the 1970s,[39] and by the early 1980s it had focused on the development of genetically modified organisms. In the next decade, it actively promoted its newest biotech invention—bovine growth hormone. Every single one of the company's lines of business have wrought disaster, and yet it still survives and thrives.

POISON APPLES AREN'T ONLY IN FAIRY TALES

Acceptance of these new chemical products has not been automatic or unanimous. Farmers resisted them. Scientists and doctors challenged them. In the 1980s, the nonprofit Natural Resources Defense Council and a group of activists raised awareness that a chemical commonly used on apples, daminozide (Alar) could cause cancer in children because they consume large amounts of apple juice and applesauce relative to their body weight.

Long before the Alar scare, many children died from eating apples contaminated with arsenic. In 1933, *100,000,000 Guinea Pigs: Dangers in Everday Foods, Drugs, and Cosmetics* was a best-selling book. The authors alleged that from the late 1800s to the 1920s, more than 100 million Americans suffered from symptoms of lead and arsenic poisoning. Lead and arsenic (lead arsenate) were used as pesticides in orchards.[40]

Believe it or not, arsenic is still used today—even in chicken feed! It's used to promote growth, kill parasites, and "improve pigmentation of chicken meat," even though arsenic is strongly linked to many types of cancer and diabetes.[41] In 1999, 318,000 pounds of arsenic were used in California alone .[42]

The US Congress responded to concerns about the food supply by establishing the Federal Trade Commission in 1912 and the Food and Drug Administration in 1927. But then as now, the industry fought standards with lobbying money and lots of advertising. The government usually sided with the groups who were making the most money—industry. Meanwhile, a host of new chemicals hit the market. Methyl bromide, a soil fumigant, was introduced in 1936, and DDT reached the market in 1945 and was widely viewed as a less-toxic substitute for lead arsenate. These are just a few notable examples among thousands. You probably have seen the pictures of trucks with hoses spraying children at play and eating sandwiches to "prove" just how safe DDT really was.

We now know it wasn't safe at all.

THE BEGINNING OF THE END
FOR SMALL FARMERS

As the freight train of the Industrial Revolution barreled at full speed through America's farm country, few small family farmers were able to survive. Artificial fertilizers did increase yields from worn-out soil, but they also increased the need for other chemicals and radically increased costs for farmers.

Many went bankrupt. Large companies bought and consolidated the farms that went under, changing the landscape of America from a patchwork of smaller, independent farms to corporate farms in a process that continues to this day.[43]

Cargill, for instance, started as a single grain storage facility in 1865. Still family owned and privately held, it is now one of the largest agricultural companies in the world.[44] While farms grew, however, the number of farmers plummeted. In the early 1900s, half of all Americans worked on farms or ranches. Today, fewer than 1 percent of Americans work on farms.[45]

It's important to realize that large-scale agriculture—which is often denounced by environmental leaders (even when the industrial-scale farms are organic)—is not a recent phenomenon. From the feudal system of the Middle Ages, when peasants farmed land in exchange for protection, to the tenant farm system, to plantations that "required" slave labor to function, our history is filled with industrial farming. America was settled by many indentured servants who had come from England to harvest the abundant resources of the new land. George Washington, through his work with the Ohio Company, was one of the many who made their fortunes by acquiring large tracts of land through "Royal Charter," clearing it and sending the timber and furs back to England, then selling it off as farmland.[46]

We believe that the past is the past and that our government is protecting us and keeping us safe. The truth is, however, that the government is a lagging indicator of safety, not a leading indicator. It responds after problems are discovered rather than acting as a shield against them. And we will be living with the impacts of our chemical hubris for centuries to come. Even today, we see our government responding to crises by setting up additional agencies, creating new overseeing roles, and throwing billions of dollars at a problem that needs to be addressed in a much deeper, more thoughtful way.

Time, however, is running out.

5. HOW INDUSTRY AND THE GOVERNMENT HAVE BETRAYED US

To make the bold changes that we all know deep in our hearts need to be made will require independent thinking and focused action. We are up against strong forces. Over decades of investing in deadly businesses, the chemical companies have mastered the art of infiltrating the minds of the public and government and anesthetizing us to any negatives associated with their products.

How do they do it?

People in general—students and professors are no different—tend to gravitate to where the money is. And the money today is in biotechnology. The industry has spent millions—perhaps even billions—to advance its interests. The chemical companies start working on the perceptions of potential consumers when they are very young by becoming major funders of organizations like the 4-H Club and Future Farmers of America. By the time those kids are in college, they are grateful for the money Monsanto and others provide for research to help them "feed the world." Moreover, they are the primary sources of funding for agricultural research at the university level. They fund research at 75 of the 76 land-grant universities (not including the 29 Native American colleges).

For example, in one year alone Monsanto provided Mississippi State University's plant and soil sciences department with $16,000 in "grant-in-aid," $12,000 for "unrestricted research," and $880,000 for "Assessing Long-Term Viability of Roundup Ready Technology as a Foundation for Cropping Systems."[1] You won't be surprised to learn that the research this money supports regularly comes out in favor of the funder.

The chemical industry's iron grip on agricultural research doesn't stop there. Genetically modified organism (GMO) seeds are so strictly controlled by the companies developing them that scientists are not allowed to study them independently. They must get permission from the companies to conduct the research, and their research must be approved by the company before it is published or submitted to the EPA. Companies such as Syngenta simply do not allow scientists to compare the effectiveness of their products to those of competitors. In 2009, 26 scientists, who declined to be named for fear of funding recriminations, submitted a statement to the EPA protesting this situation.

"Technology/stewardship agreements required for the purchase of genetically modified seed explicitly prohibit research," the scientists wrote. "These agreements inhibit public scientists from pursuing their mandated role on behalf of the public good unless the research is approved by industry. As a result of restricted access, no truly independent research can be legally conducted on many critical questions regarding the technology, its performance, its management implications, IRM [insect resistance management] and its interactions with insect biology. Consequently, data flowing to an EPA Scientific Advisory Panel from the public sector is unduly limited."[2]

Notice that there is no mention of human health. No one is required to test for the impact on human health before these products go to market. Over the past decades, public universities have come to rely on private companies rather than the government for almost all of their funding. As a result, the universities are increasingly beholden to the companies' agendas.

When universities are able to conduct research on GMOs, it's usually a fixed fight. The research plots do not include a best-practice organic plot, they only compare a GMO plot, for example, against a "conventional" plot that has not been treated with chemicals. So all of the soil is degraded and inferior in every way. Only close examination of the study reports makes this bias apparent.

Many of these chemical companies have made sure that money for research on organic agriculture is hard to come by. And when universities are forward-thinking enough to establish sustainable agriculture departments, before long one or more of the agrichemical companies becomes a major funder. In total, the chemical companies have billions and billions of dollars invested in maintaining this belief that their methods are better. It's more of a fantasy than a belief, really. But it's a fantasy most Americans have bought into.

LOBBYISTS LEAD THE WAY

Billions of dollars have also been spent on lobbying and establishing organizations such as the Alliance for Abundant Food and Energy,[3] which promotes biofuels.

The corporations and lobbying groups have actually trade-marked the word "sustainable" as it refers to GMO seeds. They try to lead public opinion by co-opting the language and arguments of the opposition. For instance, a "first-of-its-kind report" creating a framework for measuring agricultural sustainability was developed to improve agricultural production. "The Environmental Resource indicators report was released at the American Farm Bureau Federation annual meeting by Field to Market, the Keystone Alliance for Sustainable Agriculture," reads a press release promoting this supposedly great work. When you get to the last paragraph, you see who is involved.

"Field to Market members include: American Farm Bureau Federation; American Soybean Association; Bayer CropScience; Bunge; Cargill, Incorporated; ConAgra Foods; Conservation International; Cotton Incorporated; DuPont; Fleishman-Hillard; General Mills; Grocery Manufacturers Association; John Deere; Manomet Center for Conservation Sciences; Mars, Incorporated; Monsanto Company; National Association of Conservation Districts; National Association of Wheat Growers; National Corn Growers Association; National Cotton Council of America; National Potato Council; Syngenta; The Coca-Cola Company; The Fertilizer Institute; Kellogg Company; The Nature Conservancy; United Soybean Board; and World Wildlife Fund."[4]

Why are the Nature Conservancy and World Wildlife Fund on that list? I have no idea. But the American Farm Bureau is at the core of the rotten apple of lobbying. A quick check of its

Web site shows its leaders don't believe in climate change or any legislation associated with it, and they discredit any science that poses threats to the status quo of chemical farming. No wonder farmers are confused.

Lobbiers promote these companies' products and ideas to Congress, the public, and Cooperative Extension agents so that when farmers need help, the help comes from the same sources that speak with a unified voice.

A unified voice is precisely what the environmental/ organic movement lacks. When foodies sing the praises of local food sources instead of organic ones, chemical companies are laughing all the way to the bank. When I mentioned "organic" to one well-known television news journalist, she responded, "Organic? I thought that was over. It's all about local now." She's not alone in that belief.

If you go to Monsanto's or Syngenta's Web site, the picture looks rosy. Profits are up. Before-and-after photos of crops around the world show how people have benefited from their products. Their foundations are flush with cash and use it generously to fund more research and provide better seeds and chemicals to needy people around the world. Their positions seem confident and certain.

You don't hear much about Syngenta, which was founded in 2000 when the pharmaceutical companies Novartis and AstraZeneca spun off their seed and agriculture businesses to separate themselves from the negative publicity about sales of GMO seeds.[5] The histories of many of these corporations are a series of mergers, sales, and subsidiary establishings that seem designed to help the companies evade responsibility for their actions and keep customers slightly confused. And for good measure these companies make generous contributions to the election campaigns of politicians who serve on agricultural committees.

THE POWER OF DOUBT

Why do we make the same mistakes over and over again? Why do we let businesses profit from our own destruction? Why do we keep believing that we can control nature, even as it banishes us repeatedly from our homes in search of new fertile ground? Like the movie *Groundhog Day,* we are made to relive the same story over and over until we get it right.

I am not advocating going "back to nature" or turning back the clock. While I can be sentimental about the past, I hold no illusions that it was a better place and time. We need a new model. There is just one problem. We are educating ourselves out of the capacity to find solutions. We are really good at breaking things up into ever tinier bits and studying them as if they hold the secret to the universe. Meanwhile, we are not so good at looking at the whole picture—the systems approach. Whether you are talking about doctors looking at the whole body or scientists looking at the whole environment, very few have been encouraged to look at the whole picture and see how it all works together. Our educational and financial institutions reward and celebrate reductionism. Even our media are organized into neatly divided beats: One reporter covers food and another the

environment. A recent study by researchers at Johns Hopkins
University showed that stories on global warming have completely missed the food and agriculture connection, with only
2.4 percent of articles mentioning farming practices as a
contributor. Less than 1 percent made the connection between
livestock and meat production and climate change.[6]

Corporations use this reductionist approach to their advantage by instilling doubt about potentially damaging findings. If
scientists study laboratory animals, companies issue statements
claiming that animal studies do not correlate to human studies.
Then they do all they can to discredit studies in humans and, if
that fails, they discredit the scientists who conducted the
studies. When that doesn't work, the purveyors of chemicals
and GMOs lie by issuing important-looking documents, brochures, or posts on their Web sites, all accompanied by pictures
of happy children, beautiful cornfields, and patriotic farmers.
They proclaim that there is no "proof" of any danger to consumers, that their products are safe and reliable. After all, the
government has approved their products, so they must be safe.
They insist that you can't feed the world without their products
because the population is growing so fast and the available
arable land is shrinking. They want us to believe that we are all
in big trouble without chemicals! They are so effective at
conveying this message that you don't know what to think.
Doubt makes us keep our mouths shut. No one wants to be
wrong. And we are all afraid of looking stupid. The confusion
that doubt creates has been deliberate. And it works.

If you have seen the ads that Monsanto has placed in

newspapers such as the *New York Times*, you might think that
without biotechnology, drought will decimate all of our crops
and destroy our ability to eat and protect our families. Fortunately, they say, the researchers at Monsanto are hard at work
figuring out a way to engineer seeds that are more drought-
proof. With the coming climate crisis, scientists are predicting
more droughts and more extreme weather, which are great for
news ratings, but horrible for food production. The power of
advertising is being used to confuse people and distract us
from the real issues.

Milk is a powerful example. Most dairy farmers—large and
small, organic or chemical—sell their milk to a cooperative or a
company that processes (pasteurizes and homogenizes),
packages, markets, and distributes it to retailers. The price the
farmers earn depends on supply and demand. If there is too
much milk, the price goes down.

Many chemical dairy farmers rely on rBGH (recombinant,
or genetically engineered, bovine growth hormone) to
increase production, because they fundamentally believe both
that if they can get more milk out of their cows, they will
make more money, and that they need to produce more milk
because worldwide demand is increasing. They need to feed
the world.

But rBGH is not the panacea the farmers believed it to be.[7]
It causes cows to suffer a variety of problems and infections.
So the cows are treated with antibiotics. Several years ago,
Monsanto lobbied vigorously, at the state as well as the federal
level, to prohibit the labeling of any milk as "hormone-free."

(Certified organic dairy farms may not use artificial hormones.) Monsanto argued that because rBGH has not been
deemed unsafe, it confused consumers to allow labels that
mentioned hormones because they implied that hormone-free
milk is different or better.[8] Monsanto gave up on this weak case
and sold its rBGH division in late 2008.

Even if rBGH were not harmful to dairy cows or people, we
don't need more milk. We already have too much milk. So why
is everyone surprised when the price of milk drops, as it did in
March 2009, by more than 50 percent?

An article published in 2009 in the *New York Times*[9] implied
that organic dairy farmers were regretting their choice to work
within the organic standards. The economic downturn and
oversupply of milk affected organic dairy farmers, too. The
article suggested that organic dairy farmers were suffering
the most.

Organic dairyman Craig Russell of Vermont lost his milk
contract and was interviewed for the article. He complained
that his remarks were distorted on the ODairy e-mail group.

"What was not represented in the article was that I stated
that farms with little to no debt that [had] transitioned
appeared to be doing fine. I would also be doing ok if I had no
debt. I think that a rebuttal or correction is in order as I do not
represent all organic farmers' situations."

Though Russell repeatedly told the reporter he felt very
lucky that he had made the switch to organic and had no
regrets about his decision, he was portrayed as feeling the
opposite.

"Bottom line, if there was not organic we would not be milking cows," Russell said.

Newspapers are struggling these days, too, as they suffer through the biggest advertising shortfall in history. The *New York Times* is facing unprecedented threats to its survival. One company that still regularly buys full-page, full-color ads: Monsanto.

WE HAVE TOO MUCH FOOD

In 1898, the British scientist Sir William Crookes stood in front of a distinguished crowd and made a speech that gave future chemical companies the advertising sales hook they needed for the next 100-plus years. He proclaimed that unless new sources of nitrogen were found, the world—or at least *his* white-skinned, male-dominated world—would face starvation. Nitrogen, he said, was "vital to the progress of civilised humanity. . . . Unless we can class it among certainties to come, the great Caucasian race will cease to be foremost in the world, and will be squeezed out of existence by the races to whom wheaten bread is not the staff of life."[10]

Fast-forward to the 1950s. Still sounding the alarm for a looming food crisis that had yet to materialize, a promotional brochure for DuPont echoed Crookes' message:

"There is no new fertile land to break to the plow; little new crop land has been added to our farm land since 1920. Yet there is a booming rise in population. By 1975—a single generation away—another 60,000,000 hearty appetites will demand satisfaction. Unless something is done quickly, the US will have to undergo what amounts to a revolution in its eating habits."[11]

The chemical farmers I spoke to truly believe they need to increase production because they are on a patriotic mission to feed the world.

The chemical companies try to take full credit for improving yields on less land and "feeding the world," but food packaging and storage probably have as much to do with the increased availability of food. Refrigerators and freezers weren't widely available to home owners until the 1930s.[12] Before then, foods had to be smoked, dried, fermented, canned (and that was rare), or eaten right away. One reason people began to eat white rice and white flour is because when the highly nutritious brown hull was intact, the rice and flour spoiled much more quickly. You had to eat mostly local foods in season. When distribution and refrigeration made more options available, people jumped at the chance for greater diversity in their diets.

Though obesity has become a worldwide problem, chemical and biotech companies still claim that there is not enough food to feed the world—and that we *need* chemicals and GMOs to provide enough for all. They spend billions of dollars each year on advertising and lobbying in order to drive that point home. Yet the problem isn't food scarcity—it's too much food—but fear of famine sure sells chemicals.

Our ability to feed ourselves is less about production ability and more about politics. Amartya Sen, an economist then at Trinity College of Cambridge University, won the

Nobel Prize in 1998 for his study of famines. "Mr. Sen's major finding is striking," wrote Charles Wheelen in his book *Naked Economics*. "The world's worst famines are not caused by crop failure; they are caused by faulty political systems that prevent the market from correcting itself. Relatively minor agricultural disturbances become catastrophes because imports are not allowed, or prices are not allowed to rise, or farmers are not allowed to grow alternative crops, or politics in some other way interferes with the market's normal ability to correct itself."[13]

The recent global recession greatly increased hunger around the world. A study commissioned by the United Nations concluded that the quantity of food was not the cause, the price of food and political instability were. The report stressed the link between hunger and instability and reported that soaring prices (as much as 24 percent higher than the prior year) led to riots in more than 30 countries in 2008.[14]

Prices rose so fast for basic food staples around the world because of the demand for biofuel.

While I was writing this book, the domestic price of gasoline varied from $4.09 per gallon (still less than Europeans have paid for years) to less than $2.00. Adjusted for inflation, we are paying only a few cents more than we were in the early 1980s, when gas prices skyrocketed to about $1.35 per gallon, or about $3.17 in 2009 dollars. (We all heard about how the CEOs of the big three automakers, who for decades resisted improving cars' fuel efficiency, flew to Washington in three separate private jets to ask for bailouts for their nearly bankrupt companies.) When the price of gas goes down, Americans start buying trucks and SUVs again.[15]

How does the price of gas affect the price of food? It takes fuel to ship food around the world. And nearly every chemical fertilizer today is petroleum based. Most of all, our hunger for energy has convinced farmers to grow food to make into fuel.

THE GREAT BIOFUEL HOODWINK

One of the biggest problems in agriculture in recent decades has been the overproduction of corn, and when there is too much corn, its price goes down and farmers make less money. But farmers want to grow corn. It's easy to cultivate, and farmers have all the equipment to plant and harvest it. Even though output exceeds demand, the government pays them subsidies to keep growing it. So, the industry began to find other uses for this cheap and abundant raw material. High-fructose corn syrup became one use, for example, as did animal feed, even though it makes cows sick to eat corn (necessitating the overuse of antibiotics).

Ethanol is another product developed to put excess corn to profitable use. I spoke with a farmer who in the 1960s helped build the first ethanol plant. The question at the time was he told me, was "What in the hell [can] we do with all this freaking corn?"[16]

Biofuel sounds like a win-win solution: Corn as fuel could save the American farmer and reduce our reliance on foreign oil! Biofuel can also replace methyl tertiary-butyl ether

(MTBE), an additive aimed at making gasoline more environ-mentally friendly that has, in fact, poisoned thousands of wells across the country.

Biofuel from corn, however, has its own problems.

First, corn raised for fuel is even less likely to arouse concern about how it's grown than corn intended as food is. So farmers spread even more chemicals on it, thereby increasing the amount of pollution that's poisoning their families and running off their land and into streams, rivers, gulfs, and eventually oceans. In 2009, the USDA was trying to fast-track the deregulation of GMO corn crops used to make ethanol. If they succeed, this will increase exponentially our health problems, our climate crisis, and our food problems.

Second, it takes more fuel to grow corn for fuel than the corn actually produces. Making and spreading fertilizers, pesticides, and most chemicals uses a lot of fuel, and many chemicals are made from the oil and oil by-products biofuels are meant to replace.

Lastly, when food is diverted to another purpose, there is less for us to eat and prices rise. Suddenly, because corn was being grown for fuel instead of food, the price of corn around the world skyrocketed and food became too expensive for many people. Hence, the food riots.

Even former vice president Al Gore, who cast the deciding vote in favor of biofuels back in 1994, admits he was wrong about them. His extensive analysis, published in his book *Our Choice*, shows that "producing first generation ethanol from corn is a mistake."[17]

Unfortunately, the GMO "corn rush" has let the genetic beast loose and it is impossible to capture and destroy it. In Mexico, the place where corn originated and is still a major staple of the diet, heirloom varieties that have been farmed for thousands of years have been contaminated by pollen from GMO varieties (pollen doesn't respect borders or property lines). Not only has GMO corn created an unnecessary surplus of corn, but it's also ruined the clean corn that we still have.

THE SECRET HISTORY OF THE FARM BILL AND THE REAL COST OF FOOD

We Americans believe we pay too much for food. Actually, food costs as a percentage of our income were much higher in the past than they are today. In fact our food prices are artificially cheap.

In the United States we have come to expect—and even feel entitled to—certain things being cheap: gas, food, news, and anything made in China. Our primal buying habits have encouraged this belief in ways that people don't usually consider, and its effects are deeper and more harmful than we may realize. You can witness the effect of price and money in our current recession. All the righteous griping from environmentalists and people like me barely makes an impression on Americans' buying habits or driving habits, but with a financial crisis and all the fear that comes with it, shopping comes to a screeching halt and the whole economic system starts falling apart.

"Only when the tide goes out do you learn who's been

swimming naked," says investor Warren Buffett.[18] In some ways, this whole country is swimming naked. In our quest for higher-paying jobs and ever-growing corporate profits, we have outsourced production of most of our daily needs, including food, and forgotten the importance of supporting our own communities. In our quest for cheap stuff, not only have we compromised our civic and social honor and integrity, we have also eroded our national security, physical health, and independence by relying on imported goods to fulfill our most basic needs.

Farmers struggle with this conundrum on a daily basis. The American minimum wage is $7.25 per hour. In other parts of the world, millions of people live on $1 a day. American farmers find using chemicals far more cost-effective than hiring labor they may not be able to trust or who may be illegal immigrants.

We Americans like to think we are patriotic and rail about foreigners usurping our market share, but every time we go to the supermarket, we look only for the best deal we can get. So companies that make corn chips have an incentive to find cheaper corn from another country (shipping included).

Consumers bemoan the higher prices charged for organic foods. Many people view "organic" as a fashion or lifestyle choice rather than a responsible health and environmental choice, so it's no surprise that sales for organic products slow when money gets tight. Why, you might ask, are organic foods more expensive?

The answer might surprise you.

Blame it on the farm bill.

Every few years, Congress enacts legislation that sets agriculture policy for the United States. Cheap food wouldn't be possible without it. And, if not for the farm bill, organic foods would cost less than other foods. How is that possible?

The first farm bill was passed in 1933 to provide price stability for American farmers suffering through the Great Depression. At that time, deflation caused prices for farm-raised products to fall more than 50 percent, while farmers' costs decreased by only 32 percent[19] (a profit-reducing cycle that occurred again in the recession of 2008 and 2009). During the Depression, people were hungry because they couldn't afford to buy food, though farmers were producing plenty of it. They just weren't able to sell it, though, which led to even more hunger. The solution: Pay farmers *not* to grow so much with the goal of balancing supply and demand and helping them earn a better price. In 1933 alone, 6 million piglets were slaughtered.

And so it began.

The US Supreme Court invalidated the act in 1936 because the money paid out in subsidies was not being distributed for "the general good." But with the addition of a few rules for soil conservation, which Congress believed would ensure an adequate supply of foods and agricultural fibers in the future, the farm bill began its long and convoluted journey.

In 1936, President Franklin Roosevelt also launched a program to reward farmers for shifting from soil-depleting crops (corn, tobacco, wheat, and cotton) to soil-conserving crops (legumes, vegetables, and grasses). But even with a lengthy drought in 1936 and the creation of a program that paid farmers not to grow crops and instead put their land into conservation status, there was a "failure of the conservation program to bring about crop reduction as a byproduct of better land use," according to a report from the USDA's Economic Research Service.[20]

In subsequent years, farmers requested help or adjustments to subsidies from the federal government during each drought, flood, or economic crisis. At the same time, lobbyists for chemical companies demanded their employers' shares of the pie in an ongoing push-pull struggle that has led us to our problems today.

After World War II, "farmers and Government officials began to worry again that high wartime production and productivity gains from greater use of fertilizers and machinery would mean a return to surpluses and depressed prices,"[21] says the same report. More schemes were implemented to keep farmers from growing too much food. In the mid-1950s, Congress and the Dwight D. Eisenhower administration created an even farther-reaching acreage reduction program called the soil bank. By 1957, more than 21 million acres were put into reserve. But we still hadn't cut production enough.

The food surplus was so high by the 1960s that it reached crisis proportions.[22] President John F. Kennedy's attempts at reform never made it out of a congressional committee, but he was able to establish programs that directed farm surpluses to the poor and needy. He also initiated the food stamp program and extended the school lunch program to take advantage of all the extra food. By the late 1960s, the surplus problem was being solved by shipping our extra food overseas.

As chemical inputs and agricultural production continued to grow, the amount of farmland for sale also increased. Exporting food temporarily propped up prices until those bubbles burst as well, forcing the government to subsidize prices even more so farmers could continue to sell their products overseas at lower and lower prices.

In 1970, Congress authorized payments to beekeepers "who, through no fault of their own, had suffered losses of honey bees as a result of pesticide use near or adjacent to the property on which the beehives were located."[23] Sound familiar?

Exports were booming in the early 1970s, and for a single year in 1973, demand for food finally caught up with supply (a state that has not recurred since). Earl Butz was President Richard Nixon's secretary of agriculture and he encouraged farmers to plant "fence row to fence row." He also infamously stated, "Before we go back to organic agriculture, somebody is going to have to decide what 50 million people we are going to let starve."[24] Butz later resigned his post in disgrace after making vulgar racist and other similarly unacceptable remarks, but not before he had cemented the chemically dependent corn-based production system we have today.

Crop prices rose so high that the price of farmland soared, but farmers were overcapitalized. By 1978, the bubble had burst and farmers had won a $4 billion emergency loan program and a moratorium on Farmers Home Administration foreclosures.

In 1980, President Jimmy Carter suspended farm exports to the Soviet Union in retaliation for its invasion of Afghanistan. In 1982, a new acreage reduction program was again put in place as farmers were directed to cut the amount of land they cultivated by 10 to 15 percent in order to receive price supports. The PIK (for payment-in-kind) program was also developed in that year to cut production even further. Farmers signed up in droves and left idle a total of 82 million acres, the largest amount of land ever taken out of production.

"In all, enough land was removed from production to cause concern in the input industries [producers of chemical fertilizers] about the effects of PIK on purchases of supplies and equipment," said the USDA's own report. Remember, these "input industries" are the same folks who were (and still are) telling people and farmers they needed to buy their products so we can feed the world, since there isn't enough food without chemical agriculture.

The conclusion of this official 1984 report, which covers only the years 1933 to 1984 (before the GMO invasion), states that: "Price support programs have changed comparatively little in their 50 years of existence. *Price supports were designed to address the perennial problem in American agriculture—the ability of farmers to produce far more than can be consumed at home or sold abroad. As recent events show, this problem remains in part because of dramatic changes in the technology and structure of agriculture.*" (emphasis mine)

This quote reminds me of another from the DuPont Farm Chemical brochure from the 1950s.

Each American, each year, uses the output of 7.4 acres of land to fulfill his needs. A Japanese is fortunate to have for himself the fruits of less than a quarter-acre. But more land and the fertility of the soil give only part of the answer. The real difference is that the American farmer, unlike his brothers in the world's more backward areas, brings to the furrow the massive thrust of the force known as technology. It is a happy addition.

Ahhhh, the good old days. Back when the Japanese were the enemies, and before we knew they lived longer, healthier lives than we do. Back before the Internet, when people believed everything advertisers told them. Back when all the farmers were men. Wait . . . has anything changed?

THE INGLORIOUS BEGINNINGS OF SUSTAINABLE AGRICULTURE

Finally, in 1985, organic agriculture got its first fair shake and a piece of the farm bill.[25] My father, Robert Rodale, sat down with the chemical companies and lobbyists to broker a deal, and part of the deal was using the word "sustainable" instead of "organic."[26] This led to the establishment of the Low-Input Sustainable Agriculture (LISA) program. "Sustainable" let chemical companies still get a piece of the action, unlike "organic," which prohibited the use of chemical fertilizers and pesticides.

The LISA program became what is now called the SARE

(Sustainable Agriculture Research and Education) program.

"Sustainable" is a vague word that has no legal definition, which is why you can find it in every advertisement for chemical and biotech companies today. My father loathed the term (as do I), much preferring "regenerative," which implies organic because it means healing the planet and leaving the soil, the people, and the farm better than when you found it.

His fears that the word "sustainable" could easily be diluted are still being played out today. The Leonardo Academy, which has also drafted sustainability standards for the casino and gaming industry, is working with agribusiness and the American Farm Bureau to create a "sustainability" label for agriculture. Needless to say, the first draft that prohibited the use of GMOs was scrapped.[27]

The small victory of getting the LISA program funded hardly represented a major turning of the tide. In 2002, funding for the total farm bill was $273.9 billion, of which a mere $15 million went to organic research. Just $5 million was set aside for farmers who wanted to transition from chemical farming to organic, yet the organic food industry was growing at 20 percent per year and organic food processors had such a hard time meeting their production needs that they began importing much of their ingredients from other countries.

The farm bill of 2008 continues the insanity. I may be one of the very few people on the planet who has actually read the whole surprising document. First, the farm bill isn't just about farms. It covers everything from forests to conservation to school lunches (actually, that's one of the biggest items) and what we know as food stamps. Second, it's almost incomprehensible. A thorough reading requires one to have at hand a number of other major documents, such as the Food Security Act of 1985. The bill seems impossible to implement effectively—almost as if it's deliberately meant to confuse people.

The organic community celebrated its passage because funding for organic research increased to $78 million, though that is just 0.3 percent of the Farm Bill's $284 billion budget.

Now I am not saying organic farmers and researchers should necessarily get a "fair share." That sort of thinking got us into this mess in the first place. Besides, $284 billion seems like chump change when bailouts and rescue packages of $700 billion are the order of the day. But make no mistake, without the farm bill, organic food would cost less than chemical food—far less. Organic foods are already much less expensive to taxpayers. The funds spent on cleaning up the toxic messes agriculture has made of our soil, water, oceans, and health, as well as the costs of chemical foods, are impossible to calculate.

In Congress's effort to "protect jobs" (mainly at chemical companies) and American farmers, it produced a farm bill that put farmers on an economic treadmill by providing payment incentives to keep growing crops like corn and soybeans chemically and made it almost impossible to switch to organic or growing other crops. Many of the companies that make chemicals and GMO seeds, the American Farm Bureau, and the National Corn Growers Association employed lobbying firms that insisted to congresspeople that chemicals mean jobs and capitalistic freedom. The bill also enabled machinery companies to sell more expensive equipment, since so many famers will be growing the same crops the same way. We like to think we live in a free, capitalist country, but we are, actually, nothing of the sort.

Myra Goodman, cofounder of Earthbound Farms, has done the math. She and her husband sell organic fruits and vegetables grown on 33,000 acres of farmland in California (what the farm bill would call "specialty produce"). They don't own all of the land themselves. Rather, the group consists of 150 independent, certified organic farmers. They don't get a single penny from the government. In 2008, these organic farmers kept 10.5 million pounds of chemical fertilizers and 305,000 pounds of chemical pesticides out of the environment and saved 1.7 million gallons of petroleum. The carbon they have sequestered, according to the Rodale Institute's measurements, is the equivalent of taking 7,500 cars off the road every year.

Organic farmers, both large and small, are the embodiment of true American capitalism, even though they willingly abide by strict regulations to keep their farms and products certified organic.

THE TRUTH ABOUT THE FARM BILL

■ **The farm bill encourages the use of GMOs and chemicals and therefore adds many more toxins to our environment.** The chemical companies get rich, while farmers still barely make a decent living. Chemical companies are the Wall Street firms of agriculture, getting rich off of regular people looking to invest and protect their money and their futures. They sell and trade stuff that doesn't really make any sense, creating a false sense of growth and production while eating away at our vulnerable core. There won't be enough money in the world to bail them out when we finally wake up and realize what they have done to us, nor would we want to spend it on them. Right now, the chemical companies are spending the profits we've given them lobbying for more protection, more deregulation and regulations that are in their favor, and more chances to make money, and they have millions of farmers around the world addicted to their products.

■ **The farm bill lowers our export prices, putting farmers around the world out of business** and creating bad feelings about America in places where farming is often the only livelihood.

■ **The farm bill incentivizes the cultivation of foods that make us (and animals) sick and fat,** yet it bears no responsibility for the costs related to us all being sick and fat.

■ **The farm bill teaches farmers how to work a system that is dysfunctional rather than how to become better farmers, creating a sense of entitlement and co-dependency that is hard to escape.** Yet all the chemical farmers I spoke with would much rather see farm bill subsidies disappear. They look at the regulations of the farm bill as major impediments to doing their jobs. The one factor that unites the chemical and organic sides of the farming issue is their distrust of and disgust with government bureaucracy as it relates to farming.

■   **The farm bill is, without a doubt, the most unreadable document I have ever encountered.** Just a few pages is enough to shine a light on all the bureaucracy, waste, and ridiculousness of how we manage our government, food, and farms in the United States. It is more than 1,000 pages long. I had never read a bill before, but if they are all as convoluted and unclear as the farm bill, it's easy to see how we got into this mess.

Here is a sample section, chosen fairly randomly (bear with me on this, please).

7 USC 8105
SEC. 9005. BIOENERGY PROGRAM FOR ADVANCED BIOFUELS.
(a) DEFINITION OF ELIGIBLE PRODUCER.—In this section, the term "eligible producer" means a producer of advanced biofuels.
(b) PAYMENTS.—The Secretary shall make payments to eligible producers to support and ensure an expanding production of advanced biofuels.
(c) CONTRACTS.—To receive a payment, an eligible producer shall—
   (1) enter into a contract with the Secretary for production of advanced biofuels; and
   (2) submit to the Secretary such records as the Secretary may require as evidence of the production of advanced biofuels.
(d) BASIS FOR PAYMENTS.—The Secretary shall make

payments under this section to eligible producers based on—

(1) the quantity and duration of production by the eligible producer of an advanced biofuel;

(2) the net nonrenewable energy content of the advanced biofuel, if sufficient data is available, as determined by the Secretary; and

(3) other appropriate factors, as determined by the Secretary.

(e) EQUITABLE DISTRIBUTION.—The Secretary may limit the amount of payments that may be received by a single eligible producer under this section in order to distribute the total amount of funding available in an equitable manner.

(f) OTHER REQUIREMENTS.—To receive a payment under this section, an eligible producer shall meet any other requirements of Federal and State law (including regulations) applicable to the production of advanced biofuels.

(g) FUNDING.—

(1) MANDATORY FUNDING.—Of the funds of the Commodity Credit Corporation, the Secretary shall use to carry out this section, to remain available until expended—

(A) $55,000,000 for fiscal year 2009

(B) $55,000,000 for fiscal year 2010

(C) $85,000,000 for fiscal 2011; and

(D) $105,000,000 for fiscal year 2012.

(2) DISCRETIONARY FUNDING.—In addition to any other funds made available to carry out this section, there is authorized to be appropriated to carry out this section $25,000,000 for each fiscal years 2009 through 2012.

(3) LIMITATION.—Of the funds provided for each fiscal year, not more than 5 percent of the funds shall be made available to eligible producers for production at facilities with a total refining capacity exceeding 150,000,000 gallons per year.

This is just approximately 1 page of more than 1,000. Perhaps you noticed how much this Secretary has to do. He's mentioned nine times in just this section, and that's typical of the whole document. Clearly, a big staff is needed to make sure all this happens properly (the USDA, which manages the farm bill, employs 100,000 people[28]). So in addition to the $400 million that this one section allocates over the next 4 years, there are the costs of all the infrastructure and people to manage it.

I had never heard of "advanced biofuels." I am still trying to figure out the difference between biofuels and biodiesel (which is made from used vegetable oils such as french fry grease). Ethanol (the most widely known biofuel) isn't very efficient because it's not as dense as gasoline and you can't send it through the existing petroleum pipelines. So researchers and industry are working to develop "advanced" biofuels from corn, sugar, wheat, and cellulosic materials like cornstalks and switchgrass. Remember, corn, sugar, and wheat are all foods.

The United Nations reports that 75 percent of changes in commodities prices are due to diverting crops to make biofuels.[29] Our insatiable hunger for fuel (and anything, like food, that we can use to make it) jacks up the price of food around the world, so many people with low incomes can't afford to eat.

A quick Internet search on "advanced biofuels" led me to "Telling America's Story" on America.gov.com. There I learned that DuPont, the second-largest chemical company in the United States, has a "market-driven plan to generate revenue by relying on innovative research and investment in biofuels." Its three-pronged strategy includes: 1) developing new corn hybrids that will increase yields; 2) developing a new process to use the whole corn plant to make fuel; and 3) developing even more advanced biofuels. I also learned that the US Department of Energy "awarded $80 million to Broin Companies to accelerate construction of a commercial-scale biorefinery in Iowa using DuPont's new cellulosic technology." All in Iowa, of course, where farmers grow a lot of corn.

Does DuPont need the money it could qualify for under the terms of the farm bill? Thanks to the funding limitations noted in the final section of the bill, it probably can't get more than $40 million from *this* section of the farm bill. DuPont is a huge company. Its Agriculture Division's mission is to "deliver global nutrition through higher crop yields and healthier foods while developing solutions to help meet the world's energy needs." DuPont had a banner year in 2008—a devastating year for many American corporations—thanks to a 22 percent increase in sales of its Agriculture and Nutrition products, especially in "emerging markets." In 2007, the CEO's total compensation was $10,024,290—and it was a similar amount in 2008 (more, you may recall, than the cost of testing fruits and vegetables for pesticide residues for all Americans).

The federal government is paying DuPont, a highly profitable, multibillion dollar corporation, to develop a fuel made from food that will help it sell more GMO seeds and chemical fertilizers and make more money from its existing profitable food and fuel businesses. The American farmer, on the other hand, gets a tiny bit of money and a lot more chemicals (which he has to pay full price for) to feed his addiction. What do the American people get? More "cheap" food that makes us sick and fat and "cheap" fuel that contaminates our bodies, our soil, and our water and threatens our viability as a nation and as a species.

If advanced biofuels are successful, DuPont will make a fortune from them, while American corn farmers and sugarcane or sugar beet farmers and wheat farmers will still be earning pennies on the bushel.

This is just one example of how our governing process keeps chemical food prices artificially low, thereby rewarding and incentivizing bad behavior by corporations and, as a result, keeping chemical farmers stuck on that treadmill they can't get off of. American companies and organic farmers that try to do the right thing and pay their own way are not rewarded in any way for their efforts. The farm bill does not acknowledge the environmental costs, the health care costs, or the social costs of kids who are mentally, physically, and emotionally disabled as a result of our overexposure to and reliance on synthetic chemicals. It doesn't acknowledge the costs of cleaning up our water and air (which is the EPA's concern). It does not accurately reflect the real cost of real food grown by good people on land that is improving our environment rather than destroying it.

The farm bill, as we know it, must end.

GOVERNMENT AS USUAL

I recently had the honor of meeting Tom Vilsack, President Barack Obama's secretary of agriculture, at a forum on the People's Garden Initiative. Vilsack made the very bold move of digging up the front lawn of the USDA headquarters building in Washington, DC, to plant a certified organic vegetable garden— despite his connection to the biotech industry.

He told us that the programs he hopes to spearhead will be "sustainable." He plans to create a sustainable department that helps children have sustainable diets, help rural economies be sustainable, and have a sustainable presence globally. He has a true desire to see people reconnect with the land and understand where their food comes from and to make kids healthier by improving the nutritional content of foods and improving school lunches. He seems to really want to help support the growing movement to "re-supply rural America" and, as he mentioned in the meeting, the 108,000 small farmers who have just started farming in the last 5 years.

Using the word "sustainable" is safe for him because it means so many different things to people. To the chemical companies, it means better seeds and profits. To "foodies," it means local foods in season, perhaps organic. To environmentalists, it means protecting wildlife and saving energy.

But when I asked him how much flexibility he has to change the farm bill, which was approved right before he was appointed, he replied, "That's a sore subject."

To change anything in the farm bill, Vilsack must go back to the Senate's agricultural and appropriations committees, where, he says, "there's this feeling that they went through some traumatic experiences putting this farm bill together and they don't want to revisit it." While he does have some avenues for using some of the money creatively, he said he was shocked to find that in the past, money in the farm bill for rural development went to build motels.

Senator Bob Casey Jr. of Pennsylvania sits on the committee that wrote the farm bill. In an e-mail exchange with me about Vilsack's reference to the trauma of putting this farm bill together, Casey wrote,

> I think Secretary Vilsack was referring to the very long and arduous process that the Committee undertakes every 5 to 7 years when we reauthorize the Farm Bill. It is a long road to get a bill of that magnitude written in a way that balances regional interests and lines up the votes necessary for passage. Once the Committee and the Senate have completed the process and created a delicately balanced farm bill, the general attitude seems to be that we don't want to reopen portions of

that bill at the risk of jeopardizing the bigger agreements that made the bill possible in the first place. In addition, there is a very strong feeling among Committee members that the farm bill is a 5 year contract that the federal government makes with farmers. It tells them what they can expect and what they can count on so that they can make business plans and determine appropriate investments. It's important for farmers, who are really just small business owners, to have certainty with the farm bill once it passes into law.

I still wonder what he means by "bigger agreements."

While Casey acknowledges that the authority of the Committee is limitless "in theory," he also states that "the process of the Senate and composition of the Chamber where every Senator has a lot of weight in the legislative process means that the reality comes down to writing a bill that can pass first through the Committee and then through the full Senate. Obviously that means that farm bills, like most legislation in the Senate, are largely consensus bills that are fairly moderate."

In conversations with another high-level Washington insider, I questioned how much power an appointed official like Vilsack, an elected official like Casey, or even President Obama has to cut, reorganize, or radically change large government bureaucracies. The answer was not much unless one is willing to commit political career suicide.

The federal government and budget are burdened with hundreds of thousands of employees and organizations that are protected from change. Consequently, no matter which party is in power, budgets just go up and up and up. The only power an elected or appointed official has is to add, not subtract.

Senator Casey confirmed this. When I asked him what changes the agriculture committee can make to USDA staffing, already 100,000 strong, he said, "Other than the politically appointed positions at USDA, which are filled by Presidential nominations that are confirmed by the Senate, Congress has no role to play in USDA staffing. The USDA staff is made up mostly of career public servants and it is very important to keep them protected from undue political influence so that they can do their jobs. Congress can, however, make, authorize, and fund USDA to hire more people if a particular agency is considered under-staffed."

So who can cut costs and let people go, make the difficult choices that businesses, home owners, and families are often faced with when the going gets tough?

It's a mystery.

But both Republicans and Democrats have been equally responsible for creating "big government."

A reasonable American consumer might think the deck is insurmountably stacked against the organic farmer, and that there's no way to change the status quo. Think again. You have a lot more power to create the world than you think. Every choice you make creates your world. Every purchase you make rewards either the good or the bad. Every time you choose cheap, chemical food over organic, you are rewarding the bad. You vote with every dollar you spend. Companies and governments count every purchase you make and make almost every decision based on your purchases. If you buy a Coke, we make more Coke. If you buy cheap meat, we produce more cheap meat. If you keep buying big cars and continue to complain loudly about the price of gas, your friends and family will be sent overseas to fight for cheaper oil in a war we can never win. There is no "they"—businesses are guided by all of "us."

We are truly the only ones who can change the world. Only one thing can take down all the lobbyists and chemical companies and change the government. Enough of us must stand together and demand change, **demand organic**—make more noise to our elected officials than any back office relationship can withstand.

The chemical companies have spent decades and billions of dollars convincing us we can't survive without chemicals and that we need them to feed the world. It's one of the greatest financial and psychological hoaxes in history. I might even consider it a crime against humanity.

And we still have too much food.

THE GOVERNMENT IS US

The insufficient regulation of financial activities has implications not only for illegitimate practices, but also for a tendency toward overspeculation that, as Adam Smith argued, tends to grip many human beings in their breathless search for profits.

—Amartya Sen, *New York Review of Books,*
March 26, 2009

Isn't the government's job to protect its people? What about the FDA? The EPA? What is their role in all of this? What about government regulations? Didn't any of these chemicals get tested? The majority of chemical farmers I spoke with had a firm, confident belief that the products they use on their farms had been vigorously tested for safety and as long as they followed the instructions on the label, no harm would come to anyone.

Product safety testing has always been the responsibility of the company making a product, not the government. In the case of pharmaceutical products, for example, most new drugs are tested on real people in randomized controlled trials. These trials are designed and sponsored by the industry, and pharmaceutical companies pay hospitals and universities to conduct the testing.[30] The medical community has a rigorous tradition of tracking patients and their medications, allowing us to see the relationship between taking a pill and an increased risk of negative side effects like cancer or stroke. That's why so many drugs are canceled after enough people suffer negative consequences from taking their medication.

The problem with agricultural chemicals is that they are a few steps removed from the people who ingest them, so chemical companies can more easily evade accountability. Their testing on fields is often skewed from the start; their testing on human health is simply nonexistent.

And when scientists and doctors find connections between human health problems and chemicals, they often find themselves the victims of intimidation or have their funding revoked or, in a few cases, are fired.

Government action has tended, as I've already stated, to be a lagging indicator of safety rather than a leading one. In the case of agriculture, the past 100 years have seen a fairly consistent approach of populating government agencies with supporters of industry rather than supporters of safety for the people. You may have heard of the "revolving door" between chemical companies and government agencies. With each new administration, the door spins around again and fresh new chemical company supporters seem to pop up.

The introduction of GMOs into the marketplace may deserve the award for the worst of all time devious tactics that are harming the people around the world. The introduction of GMOs came during the presidency of George H. W. Bush, who focused on "deregulation." The Bill Clinton administration speeded the process along. And George W. . . . well, that's enough said. Monsanto has continuously argued that since people have been consuming DNA forever, there is a "substantial equivalency" between GMO foods and non-GMO foods.

According to Claire Hope Cummings, a lawyer for the USDA during the Carter adminstration and the author of *Uncertain Peril: The Future of Genetic Engineering of Seeds,*

The FDA uses the Food, Drug, and Cosmetic Act to review GMOs. The substantial equivalence doctrine fits nicely with FDA logic. It goes like this: any "novel" substances in food must be tested and perhaps labeled. However, if something can be "generally regarded as safe" (GRAS), as most conventional foods are, then they are exempt. Since GMOs are "substantially equivalent" to conventional food, they are considered GRAS and thus they do not require testing or labels. The EPA makes some effort to deal with the environmental impacts of GMOs. It regulates GMO pesticides (primarily the [*Bacillus thuringiensis* (Bt)] crops) under the Federal Insecticide, Fungicide, and Rodenticide Act and the Toxic Substances Control Act. The EPA operates under the assumption that Bt is safe, even though GMO Bt has been shown to have detrimental impacts on soil micro-organisms and beneficial insect populations.

In other words, there is no health testing. No environmental testing. And a long trail of suppressed research, fired scientists, and even fired government officials who questioned the approach or did tests that raised major concerns, as documented in the many books, films, and articles written about this topic. Since 1992, enough research has been done by public and private scientists to unequivocally state that GMOs and the use of the chemicals that go along with them are harmful to our health and the health of our environment. But the culture of fear that has resulted from people losing their

jobs and being publicly discredited or sued has left us vulnerable and underinformed.[31]

If every farmer switched to organic, a whole segment of our
economy would be lost. Chemical and biotech companies, lobbying firms, and the businesses and nonprofits to clean up the mess
from those businesses would all lose billions. And the government
bureaucracy that was set up to pretend to monitor all these
problems would no longer be needed, either. A lot of people have a
strong vested interest in keeping these businesses alive and
growing by any means possible. We have been manufacturing
problems and spending good money after bad to pretend to solve
them. "We make more money when we fight with each other,"
said a lobbyist I met in Washington; he works for both environmental groups and "mainline agricultural companies."

If every farm were organic, we would *save* billions of dollars
on health care, environmental cleanup, and energy. More important, we would have the opportunity to creatively make and spend
all that money in ways that do not harm people or the planet.

MONEY IS THE GREAT MOTIVATOR

The most significant and fundamental barrier to finding
solutions to our global environmental and health problems is
the belief that everything we do must make money. Hedge fund
managers and others on Wall Street are flocking to invest in
agribusiness companies, because they seem to be among the
few that are growing.[32]

If the US government truly stood up for the American

people, it would give consumers the freedom to decide for themselves what to eat and what not to eat by allowing food companies and food producers the independence to decide to inform consumers whether there are GMO ingredients in their foods. We currently do not have that freedom. If chemical companies are free to introduce risky, untested products into the market, companies like Creekstone Farms Premium Beef should be allowed to tell customers that it tests every slaughtered cow for mad cow disease. But not only did the Bush administration prohibit Creekstone from making that claim, but it also isn't even permitted to test.[33]

The government has also actively prevented food companies and farmers from labeling their products as being free of GMOs. The only way to be sure that you aren't getting GMOs and the chemicals that go with them is to buy food that is certified organic. But even this is no longer a 100 percent guarantee. Not only can some organic foods include up to 5 percent non-organic ingredients, but GMOs and their promiscuous pollens are sneaking into organic fields and contaminating our food.

As I was writing this book, I felt like a detective searching for clues. During the summer of 2008, I came to the conclusion that the root cause of our current environmental and health problems as they relate to agriculture is the government's obsession with deregulation.

And then the global economic collapse occurred.

News reports accused deregulation of causing the economic collapse, and I agree. What is behind the desire to deregulate? Greed, certainly, but also an arrogant view that greed has no consequences, that we are separate from nature and the universe—above the laws of it, not a part of it.

Deregulation has allowed many agrichemical companies to "grow" fast and furiously using subprime products, techniques, and methods—based on petroleum and toxic chemicals—that cannot continue if we are to survive.

The week that the stock market fell dramatically, I attended the Clinton Global Initiative meeting in New York City. What shocked me was the utter lack of interest, concern, and conversation about food, while people around the world rioted over food prices.

What were people talking about? Energy.

It was clear from the discussion among the world's top leaders (including many from the Middle East) that the end of oil is near. The consensus is that oil not only pollutes the world, but it also finances terror. As T. Boone Pickens said, "The sun is not a member of the Arab league." This is good news for organic farming. Organic farming is not nearly as dependent on oil as chemical farming is, especially since most chemicals are made with natural gas and petroleum *and* need a lot of petroleum to be manufactured and transported.

Despite this, we in America are more concerned about how we are going to keep our cars and TVs running than we are about feeding ourselves—and the rest of the world—*properly* and preventing the looming collapse of our ability to survive as a species. Why? Because there is lots of money to be made in energy.

Our government needs to turn its policies upside-down and

start giving tax breaks to those people and companies that are benefiting people and the planet and finding positive solutions to our food, farming, energy, and climate problems. Likewise, they need to boldly tax and fine polluters. Giving corporate villains a way to buy their way out with cap-and-trade programs seems to me inefficient, just prolonging our misery and perpetuating this industrial paradigm, which clearly isn't working and needs to be changed.

If we are ever going to rebuild trust in our government, our economic structures, and our future, we must reorder our priorities. To truly solve our problems, we need a paradigm shift in how we think about our economy, our businesses, and our government. **We have to put the safety of people and the planet first and reward the businesses who serve that mission.**

I know it's possible.

Looking at how the USDA rules for organic certification came about tells a hopeful story of government action, one in which the patience and persistence of a dedicated group of individuals brought about a set of regulations that help people trust that organic truly is organic. The USDA organic seal has become a symbol that people can trust to ensure that there are no toxic chemicals in their food. It required more than 10 years of meetings in Washington by many people dedicated to creating those rules. And it takes constant vigilance to keep the USDA organic program as pure as possible.

After the release of those rules, organic foods saw annual growth rates of 15 to 20 percent right up until the economic collapse. The government, at the urging of a dedicated and concerned group of citizens, did the right thing and the marketplace rewarded their efforts.

UNITY IS POWER

The chemical companies are not the only ones trying to undermine the USDA organic rules. Many of the challenges come from within the ranks of farmers and consumers themselves. I classify the "underminers" into two groups. The first is composed of small farmers who don't want the government regulating their farms and don't want to do the paperwork needed to get certified. The second group includes well-meaning but confused consumers as well as journalists and pundits in search of some utopian, fashion-forward ideal or shocking headline.

Let's discuss the easy one first. While there are 13,000 certified organic farmers in America, according to the Organic Farming Research Foundation,[34] there are probably a few thousand more who are organic but uncertified. These small farmers primarily sell to their customers directly at farmers' markets and CSA (Community Supported Agriculture) programs, to restaurants, or all three. The certification process is not an easy one. There are papers to fill out, inspections to pass, and standards to uphold. Some simply don't want to do that sort of work. Others are unwilling or unable to pay the thousands of dollars a year to maintain their organic status. (Even the USDA staff was surprised by how difficult the organic certification process was when they decided to put a certified organic garden on their front lawn.) Some say they don't need to get certified because they are actually "beyond organic." Or perhaps they are already working from morning until night just to get the farmwork done and don't have the time.

I sympathize with these farmers. After all, for a small farmer who is selling directly to individual people, there might not be a huge benefit to getting organic certification. They certainly won't get benefits from the government in the form of subsidies or tax credits. I wish they would join the community of farmers who have declared themselves organic, not because it's a perfect word, but because there is the potential, through civic action, to make it a better word—to ensure that it continues to mean products that are free of chemicals and farming practices that jeopardize our health and the environment. I understand why they might choose not to. It takes a lot of guts and hard work to be a farmer these days. This group is the least of the problem.

The bigger problem is people within the environmental community who are suspicious of every big business and unwilling to work with the government to ensure that access to organic food is easy, affordable, and universal. They are too busy deciding which new locavore restaurant to have dinner at to recognize the fact that the majority of Americans buy food at Wal-Mart. And that if Wal-Mart were to make a dedicated effort to become more environmentally friendly and organic (as they have, thankfully), it could do more to improve the environment than any government agency in the world could, let alone any book or chef. In a quest for the next new thing that will solve all of our problems, environmental groups and their leaders run from fad to fad. Buy local! Be vegetarian! Go vegan! Buy fair trade food! Join Slow Food! Many of the top environmental leaders have turned away from supporting Whole Foods (because it's not perfect enough) or big farms like Earthbound Farms—which have brought organic foods to millions of people—and exposed the division and lack of unity that make chemical companies positively gleeful and the government's job a hell of a lot easier. After all, small divided groups are nothing the government needs to worry about.

I interviewed John Mackey, CEO and cofounder of Whole Foods, for this book, and after a long conversation, he said of being a pioneer in the organic and natural foods world, "It's like I've been out in the wilderness hacking away with a machete for 30 years—and they drive up in an SUV and say, 'Is this as far as you've gotten?' Get out of the car and help me build the road!"

Sometimes I think the people who complain about Whole Foods don't remember what it was like *before* Whole Foods. Back then, the only source of organic foods was at a local co-op with inconsistent quality and odd hours or at health food stores where products were often rancid and moth-filled on arrival. I have often said that if my grandfather—and even my father, who died before he could ever go to a Whole Foods— would have walked into one, they would have broken down in tears. Even regular supermarkets were sterile warehouses where the only lettuce was iceberg and tomatoes were more like red baseballs. Whole Foods changed *everything* for the better. I remember. And I don't ever want to go back.

"It's like we are all standing in a circle and shooting at ourselves," says Myra Goodman of Earthbound Farms, repeating what she heard at an organic conference. What we really need to be doing is turning ourselves around and facing the "enemy." Like the insane witch hunts that have occurred throughout history, we have focused obsessively on the enemy within rather than uniting to work together to make change.

Alan Greene, MD, a pediatrician from California, recently embarked on a 3-year personal experiment to see if he could follow the USDA certified organic rules and eat totally organic (it takes 3 years of organic practices for a farm to become certified). It wasn't easy. On days when he was traveling and couldn't find anything organic to eat, he had to rely on snacks brought from home. What did he learn? Aside from feeling a lot more alert, getting sick less, and generally feeling a lot better, his biggest surprise was discovering that there are a lot of people in America who still don't know what organic is. "In stores and restaurants around the country I asked, 'Do you have anything organic?'" he explains. "Half the time the reply was, 'Do you mean vegetarian?'"[35]

More than a year before President Obama was elected, I heard Arianna Huffington say that we needed to end the false distinctions of left and right, red and blue. I believe that all people, whatever their politics, are multidimensional, complicated, and driven by the primal needs to keep their families safe, put food on the table, and improve their lot in life, if they can.

The overwhelming election of Barack Obama, whose brilliant line that "We are not blue states or red states, but the United States" showed that America is ready for unity, ready for change. It was truly a sign of hope. It is an unprecedented opportunity to align what is right for us and our children with how the government can be run. For the first time in a long time, it seems possible that the government might focus on protecting its people rather than its industries. President Obama's hero, Abraham Lincoln, established the USDA, which he referred to as "the people's department" because in 1862, when it was signed into law, farmers and their families made up half the population of the United States.

Only government regulation can protect us—and even then by only so much. Without responsible regulation, chemical companies—whether we're talking about Syngenta, Dow, DuPont, Mosaic, Bayer CropScience, Cargill, or Archer Daniels Midland—can't be held accountable, no matter how much cancer and harm their products cause. Over the years, Republicans and Democrats alike have prided themselves on moving toward deregulation. "We will take a weed-whacker to regulations in Washington," proclaimed Mitt Romney at the Republican Convention in 2008. Huzzah! Deregulation is touted as being good for growth, good for economies, and good for innovation.

There is just one problem with deregulation. Sometimes (or often) people, companies, and even politicians put their own gain ahead of the good of others—whether the others are their customers, their employees, their constituents, or even their country. History has shown that some people will choose to abuse the system and are willing to do anything to further their own gains. Regulation can't stop these people, but it is the only tool we have to protect us and prosecute them if we catch them. "The reality is that nobody ever likes the umpire," says Charles Wheelan in his book *Naked Economics*, "but you can't play the World Series without one."[36]

NATURE WILL ALWAYS WIN IN THE END

When you look at nature, you can't help but see cycles, and it's the same with economic history. There are cycles of growth and cycles of recovery (or "recession"). We are always shocked and surprised when we can't sustain perpetual growth. But is the measure of our success really growth? Or is it security? Instead of using the gross domestic product to measure the health of our economy, perhaps we should be looking at the things that define our strength and security. Things like government efficiency, employment, debt ratios, longevity, education levels, environmental cleanliness, infant mortality, and poverty levels seem like much better indicators of health and success than just the "revenue" we generate. Any good businessperson knows that gross doesn't mean much unless you know what the net is—and net can only take you so far without knowing what your debt is and if you can handle it.

We are in the throes of psychological as well as economic changes. This kind of epic change is hard. But how we respond will define us for the next 100 or even 1,000 years.

It's comforting to hang out with friends, eat delicious meals made of locally grown foods, and think we are changing the world. And sometimes that does create change. But true change is often a hard and painful process—a dirty job. People get hurt, jobs are lost, and people feel adrift for a while.

My husband and I, along with a few million other people, went to Washington to watch President Obama's inauguration, and we heard him vow to make good on his promise of change. We felt then that he would be capable of the tough love necessary to create change. But as the enormous crowds thinned and I looked around at the National Mall, I saw more trash than I had ever seen in my life—newspapers, soda bottles, hand and foot warmers, snack wrappers, and plastic bags covered the lawn. I realized then that we *all* have to change, not just the government. Salvaging our global food supply, restoring the soil's ability to sequester carbon, and removing all the toxic chemicals from the marketplace is a job of epic proportions. But we can do it.

The government needs to do its part, too. Setting up a brand new bureaucracy isn't the answer. It's time to do the right things. And the right things for the government to do are to reward and create incentives for those who are keeping people and the planet healthy and shut down or tax the hell out of those who are polluting it.

We like to believe that the government is something separate from us, some mythical "other" that causes all our problems. But the truth is, the government is us. It's made up of people, people like you and me. People who have strong opinions and convictions. People who are driven by self-interest. We are lucky that we still live in a democracy. We are still independent and free. But we owe it to our children to put aside some of our pet concerns and get involved for the greater good.

I believe we are in the final death throes of our industrial paradigm that puts business before people and the planet and considers nature an enemy that must be conquered.

We humans created the problem. And only we can solve it. We have to. Otherwise, nature will solve it for us.

PART 3

THE AGE OF HEALING

The history of food is the history of farming, and the history of farming is the history of civilization.

—Lord Northbourne,
Look to the Land, 1940

6. ORGANIC
FARMING TODAY

If we plan to stay on this planet, we must take a stand. And the very ground we stand on holds the magic key to our future here. We have the once-in-a-lifetime—no, once-in-a-species'-history—opportunity to either do the right thing or continue down the path to our demise.

But what does the right thing look like and feel like? Most chemical farmers think organic farming means nothing more than going back, returning to the way things were done before, in the "olden days"—and the chemical companies would like nothing more than for everyone to believe that. But as the chemical paradigm is about controlling nature, the organic paradigm respects nature. Business and industry are welcome in both paradigms, but they have a much more positive impact in the organic paradigm. The difference between these two ways of thinking about the universe can be characterized as disconnectedness versus connectedness. If all things are just chemical compounds that can be broken apart, controlled, and replicated in a factory, then everything is disconnected. If you instead start with the idea that everything is connected, then you can see that what you do to one part will affect the whole. Lord Northbourne explained it best in *Look to the Land*, the book in which he coined the term "organic" as we use it today.

> The best can only spring from that kind of biological completeness which has been called wholeness. If it is to be attained, the farm itself must have a biological completeness; it must be a living entity, it must be a unit which has within itself a balanced organic life. Every branch of work is interlocked with all others. The cycle of conversion of vegetable products through the animal into manure and back to vegetable is of great complexity and highly sensitive, especially over long periods, to any disturbance of its proper balance. The penalty for failure to maintain this balance is, in the long run, a progressive impoverishment of the soil. Real fertility can only be built up gradually under a system appropriate to the conditions on each particular farm, and by adherence to the essentials of that system, whatever they may be in each case, over long periods.[1]

This passage was first published in 1940, and it is even more true today, when the results of our experiments with synthetic chemistry have played out in such a negative way.

But even today, it takes courage, independence, and intelligence to become an organic farmer. You have to be willing to put up with the derision of many of your peers. You have to be smart enough to figure out solutions to your own problems, since you don't have free "crop consultants" to rely on. And in a way you have to have faith—faith that nature knows what it's doing and you are just there to help it along.

A YEAR IN THE LIFE
OF AN ORGANIC FARMER

Organic farming practices are in use on approximately 4 million acres in the United States and 30.4 million acres globally.[2] It is unlikely that any organic farmers are growing vast acreages of commodity crops such as corn and soybeans, but there are many successful large organic farms.

Like chemical farmers, organic growers also start with a seed. Or, rather, many different kinds of seeds, because most organic farmers have learned that raising a diverse array of plants (and animals) is much more effective and efficient than growing just one crop. The organic seeds they use are purchased either from independent and (when possible) organic seed sources or they've been carefully—lovingly, even—collected from the previous year's crops. After all, seeds are an organic farmer's investment in the future.

The organic farmer chooses her crops based on what she[3] knows will grow well in her climate and region, what her customers want, and probably what she loves to grow, because farming is not a life you choose to make tons of money. Organic farming is a *decent* living because you make enough money, and you feel pretty good about doing it.

She plants her seeds in soil that is rich and healthy from
years of good care. She might use a tractor to plant her seeds,
or, if she is forward-thinking, a "roller-crimper," or cover-crop
roller,[4] but she avoids driving machines on her soil as much as
possible since that compacts the soil, causing runoff and
erosion. This saves fuel as well as time.

Her tractor may not be a flashy new model, but she would
rather keep it than take on any debt. In her line of work, debt
can make the difference between profit and loss. She uses her
tractor for all sorts of things around the farm, and if her tractor
can't do a job, she pays a guy down the street who has a bigger
tractor to come in by the hour.

She doesn't need to buy synthetic fertilizers since every
few years she treats her fields to an application of compost,[5]
nature's fertilizer. She might buy additional compost if she
doesn't have enough, or she could trade with other farmers
or even pick it up at a municipal recycling center, although
she has to be careful not to inadvertently add any toxins
since her farm will be inspected every year by a government-
sponsored certifier to ensure that her fields are truly
organic.

Depending on the weather, she may or may not need to
water her crops. Her soil is spongy and holds a lot of moisture.
If she does irrigate, she can use less water than chemical
farmers have to, since the soil retains more water. The water
that does run off is less polluted than that from her chemical
neighbors' fields. To further prevent runoff and erosion,
she plants windbreaks (which the birds and bees also love)

and adds swales[6] to her fields to catch the valuable soil that runs off.

Knowing nature as well as she does, she knows that weeds will like her good soil just as much as her crops do, so she applies mulch to prevent the weeds from growing. This might take the form of a plastic sheet, straw, or leaves. If she is a particularly forward-thinking organic farmer, she planted a cover crop that is now compressed into mulch.

But weeds do grow, which is why she hires people to help her weed and employs other forms of manual cultivation to keep them under control. She tries not to till the soil to either plant or weed because she knows that doing so leads to erosion and compresses the soil. She also remembers her grandmother's saying "One year's seed, seven years' weed," so she knows that intensive weeding and killing of the weeds' seeds now leads to less weeding over the years.

She may or may not purchase crop insurance, since as an organic farmer she has to pay a 5 percent premium to get the same coverage as her neighbors who use chemicals.[7] The people at insurance companies have not read the studies that show how organic crops outperform chemical crops over time—especially in bad weather. Hmmm. Maybe she'll write a letter.

She checks for pest and disease problems on a regular basis. Since she is more likely to have a diverse group of crops, she tends to have less catastrophic problems, but every farmer has to deal with harsh weather, diseases, and insects now and then. Fortunately, the birds who flock to her farm keep the insect pests under control. But if they aren't on top of the problem, she might bring in some beneficial insects to eat the problem insects. If her crops are diseased, she tries to understand what is out of balance and correct it, either with a mineral supplement or a USDA approved organic remedy.

There will most likely be a flood or drought—organic farmers have the same weather as chemical farmers—but it will affect them differently. A flood may cause some erosion and runoff. But because organic crops have bigger, deeper, and stronger root systems (because the soil is healthier), the crops are more resilient, the plants and the soil absorb more water, and overall there is less damage and less crop loss. In a drought, the added strength of the plants' roots and sponginess of the soil make the crops more resilient and able to survive water deprivation longer. (Maybe she doesn't need that insurance after all!)

Now it's time to talk about sex again. The bees and butterflies are all over the farm, pollinating and helping that procreation along. She is concerned about colony collapse disorder in the bee community, but her bees seem to be doing all right. They are, as the song by the indie rock band Plants and Animals goes, "working hard, but hardly working." Even if there's a flood followed by a drought, it will probably be a good year. And all her animal babies are growing up fast, too. The chickens are laying, the piglets are fattening up, and the cows are producing just enough milk to make the best cheese and ice cream ever.

At some point during the year, an organic inspector will

come to make sure that she is following the rules, so people who buy organic will know it truly is organic. Keeping up her certification is a hassle and costs her a few thousand bucks per year, but it gives her the right to charge enough for her crops to cover her costs with a little bit extra left over. She has no government subsidies to fall back on. The government doesn't "bail out" organic farmers. But she also has more freedom and sleeps well at night. Nor does she have to deal with the "gene police."

Her harvest is not a single event, but rather many smaller ones, so her risk is spread out over the year. Since there has been such high consumer demand for organic foods, she has a few more choices than chemical farmers do of where and how to sell her crops. She might have formed a CSA (community-supported agriculture) group that allows her friends and neighbors to buy her goods in advance and receive a weekly distribution in return. Maybe they even help her do the harvesting! She might have a stand at the local farmers' market, where she can look her customers directly in the eye as she sells the labors of her love. A local restaurant might feature her products on its menu, which helps her win credibility in the community. If she is certified organic, rather than an "underground practitioner," she might have a contract with a food processor like Stonyfield Yogurt (owned by Groupe Danone) or Cascadian Farm (General Mills), a dairy co-op like Organic Valley, or a direct relationship with a supermarket like Whole Foods, Wal-Mart, or my local supermarket, Wegmans. There is virtually no chance that her corn will be used for biofuel, since the varieties she grows for food are much more valuable.

She might rely on migrant workers or labor contractors for harvest every year, because they are the only ones willing to do the work. But she tries to treat them well and looks forward to seeing families return every year. Fortunately, many young people are getting interested in organic farming and offering her sweat equity in exchange for the chance to learn how to farm. She sometimes puts her kids to work, too—they love collecting eggs and feeding the animals. She feels good knowing that they won't be exposed to any dangerous chemicals or toxins on her farm.

And now for the food—the glorious food. Her products, because organic is less available and the demand is high, go to people who really care about what they eat—and, yes, who can afford the higher cost. Some may go to a food processor, and although fuel is required to process and transport it, most organic processors know that their customers are concerned about the environment, so they are careful to use less, and less toxic, packaging.

After the harvesting and the harvest feasts and festivals are finished, she still has two more jobs to do. The first is to turn her farm waste into fertilizer. She gathers the stalks and leftover plant materials, the animal manure and bedding straw, and adds them to her compost pile, which "cooks" as its materials decay over the winter and turns into fertile soil that will enhance her farm's health and productivity.

Last, she plants a cover crop. She started doing it to cut

down on winter erosion and weeds—which it did well. Then she read an article about mycorrhizal fungi, which grow on the roots of her winter cover crop and pull carbon from the air to hold it safe and sound in her soil. Though she can't see the mycorrhizal fungi, she knows they are there, working hard for her farm. They are on her team, just like the birds, the bees, the chickens, and the flowers.

Organic farming is a hard life, but it's a good life.

At the end of the organic cycle, no corporation has made a lot of money selling stuff to her. Organic farmers aren't held hostage either. No government agency has to clean up after her, but the government hasn't helped her either (other than by establishing those USDA Organic rules and not allowing them to be eroded—one case in which government regulation benefits us all). No river or ocean has been polluted. No child has been poisoned. More carbon has been stored on that farm than she released in the process of farming her land. Best of all, healthy, nutritious, and tasty food has been provided to the people who were able to buy it from her. And she's made the world a better place in the process.

That sounds good to me.

THE ORGANIC PIONEERS

Since the beginning of the human synthetic-chemical experiment, a few important people have not only questioned the use of chemicals, but also explored alternative ways to grow food—often based on historically effective techniques that had been lost or overlooked due to the arrogance of "civilized man."

In 1911, F. H. King published *Farmers of Forty Centuries*, which explored how China was able to grow highly productive crops on the same land for thousands of years by using human and animal fertilizer. In 1924, Rudolf Steiner gave eight lectures and later published a book called *Agriculture*, which became the manual for biodynamic farming. Sir Albert Howard's book, *An Agricultural Testament* was published in 1940 and introduced the Western world to the techniques of composting that India had used for centuries. In 1943, Lady Eve Balfour wrote the book *The Living Soil,* which recognized the importance of soil health to growing food and human health. She founded the Soil Association in the United Kingdom in 1946.

In 1940, about the time that my grandfather, J. I. Rodale, was moving to a worn-out farm in Pennsylvania to test his ideas for organic farming, Lord Northbourne wrote *Look to the Land.* In it, he writes:

> In order to deal with the loss of fertility, or to enable speeding up to be accomplished without incurring loss, scientists have lately come to the rescue with a great variety of "artificial manures." By their use remarkable immediate results can often be secured in the crops to which they are applied. But it may be that we have made the same kind of mistakes in feeding our land as we have made in feeding ourselves, and that most of our artificial manures must be regarded at best as stimulants rather

than as foods, and in no case therefore as substitutes for biologically sound feeding.[8]

When my grandfather launched *Organic Farming and Gardening* magazine in 1942, Sir Albert Howard was a contributing editor. J. I. Rodale was ahead of his time in many ways, but he was also a part of his time. His magazine served as a voice for a growing chorus of concern, and back then, if you wanted to eat food without chemicals, there was only one way to do it: Grow your own.

J. I. Rodale's ideas did not take hold immediately. Rather, he encountered outrageous ridicule and prejudice. The people developing new chemicals—intellectual descendents of those once called quacks—used that label for my grandfather in a classic case of "done-to becomes done-by." By then, most Americans did believe, thanks to shrewd marketing, that chemicals actually *made* food safe to eat and that plants wouldn't grow without them. It was downright unpatriotic to believe otherwise.

But on his little farm in Allentown, Pennsylvania, J. I. continued to experiment. When he and my grandmother had bought the farm in 1940, it was about as unfertile as a farm could be. He jokes in his book *Organic Merry-Go-Round* that when they first bought the farm, even the rats were miserable. "Any untaught amateur in the lore of rodentry could see that these specimens were the worst of their scummy race. . . . They seemed savagely displeased and snarled as they ran." But with his organic methods, the farm was soon completely

regenerated. The rats, he wrote, had become "amiable creatures that reminded me of opossums."[9]

When one of the scientific community's own, Rachel
Carson, published *Silent Spring* in 1962, the world started to
take notice of our chemical dependence. But the public and
media focused primarily on DDT, which was only one of many
toxic chemicals in use. The scientific evidence *Silent Spring*
presented did rally public opinion against DDT, which led to a
ban on its use in the United States. However, it is still in use
today in Africa and India.[10]

The community of organic supporters in America remained
small and consisted mostly of independent-minded gardeners
and conservationists until the early 1970s. Then hippies
embraced organic farming and food as an avenue for political
independence and self-sufficiency and "organic" became a
"movement." One person from that emerging counterculture,
Alice Waters, opened a restaurant in Berkeley, California, and
with her leadership, organic gradually changed from primarily a
health nut's pursuit into a gourmet culinary pursuit. At the
same time, in Pennsylvania, my father, Robert Rodale, began
planning a scientific study aimed at comparing the results of
organic farming with those achieved by chemical farming
methods (the Farming Systems Trial).

The organic idea began spreading around the world.
Masanobu Fukuoka wrote brilliantly about organic Japanese
rice farming in 1978, and Bill Mollison, a Tasmanian, developed
the principles of permaculture, which defined an organic
approach to a whole landscape.

Yet the term "organic" remained unprotected. Some states, such as California in 1979,[11] had passed their own certification laws. But not until 1990, the year of my father's sudden death in a car accident, did Congress pass the Organic Foods Production Act, which directed the USDA to develop national standards for organic products. The result is the USDA Organic label that we now see on many items in supermarkets and other stores. For 7 long, contentious years, industry leaders, farmers, consumer advocates, a few enlightened scientists, and the government worked to come to agreement on those standards.

When the first set of proposed rules was issued, the public was offered the opportunity to comment. The USDA received nearly 300,000 comments, most demanding that irradiated foods, sewage sludge, and GMOs be prohibited for certified organic products. That's more letters (e-mail was not yet in widespread use!) than the USDA had received ever before or since.

Today, the USDA Organic label is found in products in almost every supermarket in every state and demand for organic products has never been higher. Britain's Prince Charles has an organic farm and business, and is one of the leading advocates for the cause. First Lady Michelle Obama has joined with local schoolchildren to plant an organic garden on the White House lawn (although she doesn't call it "organic" publicly). And the USDA headquarters has a certified organic garden on *its* front lawn.

So we've won, right?

Wrong.

The amount of organic farmland in the United States is less than 1 percent. The amount of organic farmland in Europe is slightly more, 4 percent. But worldwide the figure is still less than 1 percent of all agricultural land. "A rounding error," as Gary Hirshberg, CEO of Stonyfield Farm, calls it. GMOs are now the rule rather than the exception, allowing herbicides like Roundup to be spread indiscriminately around the world in ever greater quantities. The impacts on our health, environment, and climate are catastrophic.

7. THE TRUTH ABOUT MODERN ORGANIC FARMING

The chemical industry would have you believe that organic farming takes a "do nothing" approach. If you've ever planted a garden, you know that doing nothing doesn't work. Doing nothing is not the foundation of organic farming—just the opposite, in fact. Thanks to the Rodale Institute's work, as well as that of many organic farmers and researchers around the world, we now have a much better understanding of how to successfully grow food organically. Not just "sustainably," but regeneratively, focusing on building soil and ecosystems that are healthier than when we began and able to heal the damage we have caused.

The Farming Systems Trial (FST) study at the Rodale Institute began in 1981 as a way to study the effects of *transitioning* a farm from chemical to organic methods. At that time, no university or business would conduct scientific research on organic farming—it was viewed as an inefficient, fringe method of farming.

In the early 1980s the plight of farmers felt dire. They were in the grip of a crisis, thanks to the Earl Butz farm policy. Farmers were growing too much food, so prices dropped sharply at the same time that the price of land dropped. Many farmers were overburdened by their debts and only the largest-scale farmers could survive. During this period, countless small family farmers were forced off their land. (The Farm Aid organization, which hosts a benefit concert series, was founded as a result of this crisis.) My father, a highly patriotic man whose value system was founded on the Jeffersonian respect for the American farmer, felt an urgent need to find solutions to these problems. He wanted to ensure that small farms in America would stay with the families that had cultivated the land for generations, and that they could earn a fair living from their work.

In the 3 decades since the FST began collecting data, the Rodale Institute has researched the best practices of organic farming and shared that information with farmers, who have trusted and applied it to cultivation on their land.

The FST has produced numerous valuable findings. Here are just a few.

■ **Crop yields from organic and synthetic-chemical farms are similar in years of average precipitation.** The idea that organic farms yield less comes from the chemical companies, who tested their products on degraded and damaged soil. *Once the soil is restored organically*, organic crop yields are comparable to the latest chemical yields.

■ **Organic farm yields are *higher* than those of chemical farms in years of drought,** due to organic plants' stronger root systems and better moisture retention in the soil. Organic soil actually *looks* like a sponge when you see it under a microscope.

■ **Organic yields are also *higher* than chemical yields in flood situations,** again due to stronger root systems and the soil's ability to absorb more water and prevent runoff and erosion.

■ **Organic production requires 30 percent less fossil fuel than chemical production** when growing corn and soybeans (the two crops with the largest shares of farmland in the United States).

■ **Labor inputs are approximately 15 percent higher in an organic farming system.** In other words, organic farming creates jobs.

■ **The net economic return for organic crops is equal to or higher than that for chemically produced crops** because upfront costs are lower and organic foods typically bring higher prices in the marketplace. And that's without government subsidies.

■ **Organically farmed soil has significantly greater carbon and nitrogen storage capacity than that of chemically treated farms.**[1] Mycorrhizal fungi that naturally occur in the soil are a sign of health, absorbing carbon from the air and storing it for decades. Nitrogen is also much more likely to stay in organic soil, where it can be utilized by the plants rather than running off and polluting groundwater.

When managed properly, nature finds a way to balance things out. When we collaborate with natural processes, an intricate chain of natural checks and balances leads to a very harmonious, beautiful, and happy environment. The natural controls go out of whack when we overmanage, overuse, and overcontrol nature. We overmanage nature when we try to grow a single crop on thousands of acres, when we try to squeeze thousands of cows into a space too small for even a few hundred, and when we try to kill just one type of bug and that throws three other bugs or the birds that feed on them into disorder.

Nature needs time to heal. Transitioning from a chemical farm to a certified organic farm takes 3 years, and during that time there may be more problems. Like a person going into rehab for substance addiction, it often feels worse before it feels better.

Organic farming is more labor-intensive than chemical farming, and therein lies an irony. The government is always seeking to create more jobs, but its actions actually emphasize the drive for "efficiency" for farmers, which means fewer human hands and more hours spent alone in giant tractors. Farmwork is hard work, there is no question about it. It's also satisfying, physical work that has sustained American families for 3 centuries. Just as important as the jobs is how we treat the people who do them. No food system can ever be good for us, says Eric Schlosser, author of *Fast Food Nation*, "if the people who harvest our food, process it, and prepare it for us are treated badly."

Switching to all organic food production is the single most critical (and most *doable*) action we can take *right now* to stop our climate crisis. Organic farming can pull, on an annual basis, thousands of pounds of carbon dioxide per acre right out of the air and keep it in the soil, adding to its carbon stores year after year. We can go from being carbon polluters straight to being carbon savers, bypassing "carbon neutral" phase. We are only beginning to understand just how powerful our soil can be.

Suspend all the chemical propaganda you've heard over the years and follow along as I explain how modern organic farming really works.

Organic farming is about more than not using chemical products. Organic farmers employ a variety of techniques to keep their fields productive and healthy.

NO-TILL FARMING

The practice of tilling the soil began back in the earliest farming days, based on the fairly logical idea that by breaking up the crust of the soil you make it easier for seeds to take root and grow and to keep weeds under control.

However, repeated tilling breaks down the structure of soil, causing erosion and runoff, and disturbs the microbes and fungi in the soil that support healthy plant growth. Studies show that tilling combined with chemicals results in little or *no* carbon sequestration.[2] Organic methods, on the other hand, can sequester lots of carbon in the soil—even with a bit of tilling. A 9-year study done by John Teasdale, PhD, a plant

physiologist at the Agricultural Research Service (ARS), which
is the USDA's chief scientific research agency, confirms the
findings of the FST. In a separate 3-year follow-up study,
"Teasdale grew corn with no-till practices on all plots to see
which ones had the most productive soils. Those turned out to
be the organic plots. They had more carbon and nitrogen and
yielded 18 percent more corn than the other plots did," according to an article in *Agricultural Research*.[3]

"Tillage is a root cause of agricultural land degradation—
one of the most serious environmental problems worldwide—
which poses a threat to food production and rural livelihoods,
particularly in poor and densely populated areas of the developing world," reported the authors of an article in *Scientific
American* magazine.[4]

Tilling impacts climate change in other ways as well. Each
time farmers operate their tractors, they compact the soil,
harm the microbes in it, and burn fuel, spewing even more
carbon dioxide and other pollutants into the air.

Masanobu Fukuoka, an organic pioneer from Japan and
author of *The One-Straw Revolution*, saw the threat of tilling
back in the 1970s. He developed a method of coating seeds in
clay so they could be sowed on *top* of the soil. "That which was
viewed as primitive and backward," he observed, "is now
unexpectedly seen to be far ahead of modern science."

Not tilling does not make a farm organic, but it is a key to
being successful at organic farming and reducing the carbon
impact and increasing its sequestration. An unintended
consequence of no-till farming is that it encourages chemical

farmers to use *more* chemicals. The conservation regulations in the farm bill now require farmers to use no-till practices for erosion control. But weeds still grow. and instead of tilling or managing weeds organically, chemical farmers add a few more rounds of herbicide[5] rather than planting cover crops or applying mulch, methods that are both more effective and better for people and the planet. Chemical farmers may plant cover crops, but then use herbicides to kill the plants before planting their cash crop. Many of the chemical farmers I spoke with felt that requiring them to use no-till practices put them in a bind that forced them into using more chemicals, and they blamed the government for the problem. The real problem lies in the approach of picking one or two techniques of organic farming and not recognizing that it works as an integrated system. That system may vary for each farm, each region, each country, and each continent. However, the fundamental principles of organic farming are the same.

COMPOST

Compost once was just a smart way for people to handle their waste. It was a virtuous cycle of producing food, consuming food, and then putting the leftovers back into the earth for the next cycle of growth.

This is how nature grows trees, for instance. In the spring, the leaves grow and the tree draws nutrition from the earth. In the fall, the fruit and leaves fall. The fruit is consumed by animals (and people), who pass it on in the form of waste. Waste is nature's ultimate fertilizer. The leaves decompose naturally over the winter and provide a fresh batch of fertilizer for the tree in the spring. It's efficient, it's effortless, and it's entirely free.

On chemical farms, the whole cycle has been smashed apart and now, like Humpty Dumpty, it's hard to put it back together again. Farmers today, and many home owners for that matter, remove and dispose of leaves and other natural wastes and then buy fertilizer. For instance, dairy farmers have a lot of cow manure. Organic dairy farmers spread composted manure on fields and manage their pastures organically. They remove carbon dioxide from the atmosphere and sequester it. Chemical dairy farmers typically do not grow their own feed or have pasture for animals to graze on, so they don't have fields where they can spread the waste. Instead, they use large giant vats to contain and process all the methane gas that concentrated piles of animal waste produce. Chemical dairy farmers *add* carbon dioxide to the atmosphere (and, worse, lots of methane). Same basic activity, much different outcome.

Organic plant waste also can and should be composted. It is just as important as animal manure to rebuilding the fertility of the soil each year. One of the drawbacks of the biofuels industry is that it diverts plant waste from use as natural fertilizer for the soil. Removing that waste from farms takes the carbon away and burns it as fuel, creating even more carbon dioxide, rather than sequestering it in the soil where it belongs.

If every farmer—make that every person—composted his or her food and yard waste (kitchen scraps, leaves, grass clippings, etc.) and added it back to their land, we might be able to stop all climate change right now.

COVER CROPS

Plants grown to nourish the soil and prevent weeds, known as cover crops, are essential elements in the carbon-sequestering capability of an organic farm. They also prevent soil erosion over the winter months, functioning as a warm, green blanket for the land. Many different types of cover crops are used in different regions and for different purposes. Some are planted in the spring specifically for weed control or to benefit later crops. Many of them are legumes, which pull nitrogen from the air to enrich the soil for the next crop.

The Farming Systems Trial found another reason to use cover crops. Our new best friends, mycorrhizal fungi, which grow on the roots of plants and have a symbiotic relationship with them, thrive on cover crops. At the same time that cover crops are protecting the winter soil from erosion, giving farmers free nitrogen from the air, and preventing the growth of weeds, they are also providing homes for carbon-eating mycorrhizal fungi.

CROP ROTATION

Synthetic-chemical farms (and the equipment it takes to farm that way) have become so big and specialized that their ability to rotate and diversify their crops is extremely limited. The most successful farms include animals, which are pastured on one area, then rotated to another—long before the pasture is over-grazed and turns to mud. These pastures are tilled once every 3 years to incorporate the animal manure, a key fertilizer, into the soil, to feed a diverse array of crops for the next few years. Rotating the crops among the fields from year to year breaks the cycle of disease and confuses the insects enough that chemical pesticides, herbicides, and fungicides are not necessary.

AN END TO CONFUSION

If the evidence that organic is better for the environment is so clear, and the research has shown that organic farming is more profitable and productive over the long term, why hasn't every farmer switched to organic methods? Especially if farming organically can also stop the climate crisis, save the finite supply of petroleum oil for other uses, and eliminate the majority of the toxins from our soil and water and thus from our bodies?

Attitudes can and must change. Chemical companies must not be allowed to exert undue influence over the agricultural research conducted on our nation's campuses, and the government must revisit and rethink the incentives they offer farmers to produce chemical crops. Farmers must work together to help each other transition to organic and become successful modern organic farmers—not only supplying the world with healthy food, but also healing the planet in the process.

But how do we begin to untangle the mess, begin to work ourselves out of our chemical dependency and into a world that truly is organic, regenerative, and, most important, healthy for all? We need to end the confusion once and for all and unite over what really matters.

Well, let this be your cocktail party guide to global organic conversion, no spin included—just the facts, and maybe a few opinions thrown in for good measure.

Chemicals are not necessary to grow food. Synthetic fertilizers, pesticides, fungicides, and GMOs are substitutes for thinking, understanding, and effort. They are necessary only to generate large profits for businesses and to dispose of our toxic industrial wastes. Virtually every food in the world has been successfully grown and made organically in modern, productive, and regenerative ways—from fine wines to white flour, apples, cherries, the most delicious gourmet beef, olive oil, and yes, even lard.

Chemicals make food, soil, water, and air dirty, toxic, and poisoned. The manufacturing, transportation, and use of chemicals for agriculture are energy-intensive and poisonous to all things that come in contact with them. Most chemicals don't biodegrade within a few months. Like nuclear waste, some toxins last forever, and many of the impacts are known to be horrible. Already, dead zones in the ocean are starting to spread, wells are contaminated, and we suffer increasingly from infections and diseases such as asthma, diabetes, methicillin-resistant *Staphylococcus aureus*, Parkinson's, and cancers associated with chemicals.

Agricultural chemicals destroy the soil's natural ability to store and process carbon, thereby leading to our gradual suffocation and a global climate crisis.

Mycorrhizal fungi in the soil are our greatest allies in the fight for our survival on this planet. They are the hidden heroes beneath us. Chemicals kill them. Chemicals kill us.

Smaller doses of chemicals can be just as dangerous as large doses. Most of the government regulations on chemicals are based on estimated safe amounts of exposure. Doctors and researchers are finding, however, that small doses, and cumulative small doses, can be just as toxic as large doses. There really are no safe limits.

Organic foods are healthier.

■ Growing foods organically prevents thousands of highly toxic chemicals from entering our environment and poisoning our soils, our wells, our wildlife, our children, and ourselves.

■ Growing foods organically restores the earth's ability to process and store carbon and thereby significantly reduces the atmospheric problems that are causing the climate crisis (whether you believe in it or not!). Even more important, restoring the earth's ability to store carbon will help us all keep breathing.

■ Studies have shown that some organic foods are higher in antioxidants and powerful cancer-fighting nutrients such as conjugated linoleic acid. And, when you eat them, you are ingesting much cleaner, purer food.

■ In one study published in *Environmental Health Perspectives,* researchers analyzed the urine samples of children when they ate a chemical-food diet and then again after eating an organic diet (before returning to a chemical diet again). The researchers documented a significant decrease in agricultural chemicals detected in their urine on the same day they

switched to an organic diet—and a return to higher levels
when they went back to a chemical diet.[6]

Organic foods are safer. No food system will be ever 100
percent safe because processing factories and home kitchens
can be unsanitary, but organic foods are safer because they are
produced without dangerous chemicals, antibiotics, and risky
(to say nothing of disgusting) but cheap practices like feeding
dead cows to living cows and applying contaminated sewage
sludge to farm fields. Certified organic products are the only
foods available that have a government-backed guarantee that
no chemicals, antibiotics, sewage sludge, or GMOs were used in
growing or processing the foods.

That said, wash *all* foods (even prepackaged organic
produce) and handle them carefully before eating them.

**Organic foods are natural, but natural foods aren't
organic.** No official standards define "natural" for producers
or consumers, but it tends to denote foods that contain no
preservatives or artificial ingredients. There is no independent
confirmation that products meet that definition—it's just a
word on label that any food producer can use. Organic foods, by
the standards defined by the USDA and independently verified
by inspectors, must be natural.

Organic foods taste better. There may not be any scientific proof to back this up, but anecdotal evidence demonstrates
that organic produce has overcome its once-negative reputation as being bland or tasting bad. Now organic foods are
thought of as fresh, vibrant, and flavorful. The bad press about
the poor taste of organic foods has its roots in two things. First,

since the "organic" movement started with the health movement, it was often combined with no-salt, no-fat, no-flavor, no-thanks cooking, which truly was less tasty. Second, when organic foods originally became available in the marketplace, farmers were just relearning how to grow without chemicals, so the fruits and vegetables were often unappealing to look at and therefore less appealing to eat. Today, organic fruits and vegetables look just as good if not better than produce from chemical farms, and they taste even better—especially if they are grown locally and purchased ripe and fresh.

In an unscientific study, my kids confirmed that organic foods taste better. And frankly, isn't that the group that matters most?

Choosing to eat organic foods does not condemn you to a diet of nuts, berries, and tofu. Today, you can find organic versions of the most popular foods, including such favorites as Hidden Valley Ranch dressing and Heinz ketchup. It is possible to produce any food organically, even Cap'n Crunch cereal and American cheese.

Eating organic doesn't make you a vegetarian. Both meat and dairy products are now produced organically. Environmental concerns about eating meat center primarily on the amount of methane produced by raising animals in concentrated animal feeding operations, which are also inhumane, wasteful, and rife with diseases, necessitating the overuse of antibiotics. Studies have shown that eating too much meat is not healthy, but farm animals are an important component of farm fertility and human nutrition.

A vegan versus meat-eating fight diverts our attention from the key goal of producing and eating only organic food.

We need to unite for the planet right now. Vegetarians and vegans *especially* need to demand organic because they often depend on soy foods for protein. Most nonorganic soy is GMO.

Organic foods aren't just for hippies. Seriously, all sorts of people eat organic foods. Poor people, rich people, liberal Democrats, evangelical Republicans, celebrities, and non-celebrities eat organic foods. The Obamas eat organic foods. Even the CEO of Monsanto, Hugh Grant, eats organic food![7]

Organic farming over time is more productive than chemical farming. During times of drought and floods, organic farms consistently produce more than chemical farms. During ideal weather, chemical farms can produce more than organic farms. But there has never been a guarantee (governmental or otherwise) of ideal weather.

We can feed the world with organic foods and farming. Despite the propaganda churned out by biotech and chemical companies, organic farming is the only way to feed the world. Transferring our toxic agricultural system to other countries is sure to bring about a global environmental collapse. The energy required, the toxicity of the chemicals, and the degradation to the soil will be fatal. Instead, we need to export the knowledge we have gained about successful modern organic farming and then help others adapt these practices to their climates, regions, and cultures.

Organic is more important than local, but local is also important. Numerous studies have shown that organic is much more critical when it comes to carbon than local. In one study commissioned by PepsiCo, an independent researcher determined that the most significant component of the carbon footprint for Tropicana orange juice (a PepsiCo product) wasn't transportation or manufacturing, but "the production and application of fertilizer"[8] required to grow oranges.

The local food movement has been very important in revitalizing small farms and communities and bringing fresh, seasonal foods to many more people. However, as a means of saving the planet and improving our health, it only goes so far. Local chemical farming contaminates local communities and actually increases residents' carbon footprints and energy use. Local *organic* farming cleans up communities and decreases their carbon footprint and energy use.

International trade is essential. Coffee will always be grown in places other than North America. Cacao, too, which is what chocolate is made from. Are we willing to give those up? No. Nor should we. The glorious and devastating history of our planet is filled with people fulfilling the primal urge to trade, to explore, to exchange—from the Silk Road to Route 66. And in those exchanges we learn. We learn about people who are different from us. Different religions. Different ways of healing. Different foods. Without trade, Italians wouldn't have pasta, Irish wouldn't have potatoes, and Americans wouldn't have pizza or sushi. (Even Alice Waters wouldn't have become the revolutionary chef without her famous trip to France, where she first tasted real food.)

But we also learn what is the same. We love the same. Our bodies, while they are different shapes and colors, all work the same. We all are born and we all die. And we all like to eat!

Fair trade is the best way to help people in other countries. There are 1.1 billion people around the world who make less than $1 a day[9] and live in desperate poverty. Their best hope for improving their plight is creating organic products to sell to other countries. And frankly, it's the best and most respectful way we can aid them.

Trade often has negative consequences—intentional and not. Weeds, rodents, diseases, and insects are just as likely to travel as humans. And our culture of fast food and violent entertainment has contaminated countries that were much healthier before we "exposed" them to our poor habits. Melamine and lead contamination have taught us that greed and corruption are universal. And now we are exporting our chemical-agriculture addiction. Trying to overcontrol trade—whether of food or information—can lead to isolationism and dangerous political climates. Undercontrolling trade can be just as devastating.

Organic farming increases and protects the planet's biodiversity. If you are an animal lover of any kind, organic is for you. A recent report by the International Union for Conservation of Nature documents that "life on earth is under serious threat." The report found that one-third of amphibians, at least one in eight birds, and a quarter of mammals are on the verge of extinction. Half of all plant groups are threatened.[10] Development and logging are responsible, but agriculture is as much, if not more, to blame. As I have explained, the toxic effects of chemicals have reduced all species' abilities to survive and reproduce.

Growing organic is not going backward. When I proposed to chemical farmers that they switch to organic methods, they frequently replied, "Do you mean going back to the old way?" No! I believe in applying the best of modern science, technology, and resources to constantly improve our understanding of nature and our ways of growing and producing food. I also believe we cannot let corporations profit from killing us.

Chemical farming eliminates jobs. Many farmers use chemicals to make their jobs "easier" by allowing them to farm more land with less labor. That means fewer jobs. Whenever you hear a politician promising more "jobs," think twice. There are lots of different kinds of jobs. Which job do you prefer: working in a coal mine or on a farm? Working in a chemical factory, knowing that you are likely to die from cancer and are contributing to causing it in others, or hard physical labor outdoors that is likely to prolong your life? We have traded good, healthy, and hard jobs for easier, unhealthy, and more deadly jobs—all in the name of "production efficiency."

Government subsidies are the primary reason for the low prices of chemical foods. Without government subsidies, chemical food would not be less expensive, but rather much more expensive. Without government subsidies, farmers would not find it cost-effective to increase the amount of land they have in cultivation, buy larger machinery, and farm more with less labor by using chemicals to artificially increase their yields. Organic foods have no hidden costs.

As a citizen, the minute you decide to stop using chemicals in your yard, you have gone organic. Chemical lawn fertilizers, herbicides, and pesticides used at home are just as dangerous, toxic, and threatening to our health and environment as chemicals used on farms. You can simply decline to use them. Home owners don't have to become certified organic because they aren't selling anything. Making the choice to eliminate chemicals from your home and life, on the other hand, is a powerful and important act that will benefit your family's health immediately. You can find many free resources to help you solve problems with your lawn and garden without chemicals. Rodale.com is a helpful place to start looking for resources.

Organic farming, gardening, lawn management, and living can stop the climate crisis—whether or not you believe it exists! When you combine the impacts of protecting the mycorrhizal fungi, eliminating all the toxic chemicals and their packaging and the amount of energy spent producing them, and reducing the amount of time you spend driving to the doctor with all the resulting health problems, the carbon problem in our atmosphere is practically solved. We still need to invest in renewable energy—but restoring the earth's ability to sequester carbon is a good place to start.

Regeneration is completely possible—we can make new soil. When my grandfather bought his farm and my father bought the "new farm," both farms were degraded, eroded, and unfertile. Both farms are now paragons of fertility and filled with healthy, thriving plants and animals. In fact, for a study at the Rodale Institute on cutworms (chemical corn's greatest pest), the researchers could not find enough cutworms and had to get some shipped in!

The USDA's organic standards are good enough, but we need to make them better. The integrity of the organic label is more important than the scale of any farm. While it's lovely to contemplate a world where each of us purchases all of our food directly from the well-intentioned person who grew it, this simply is not realistic. The majority of the millions of people in America are shopping in supermarkets, Wal-Marts, gas station mini-marts, and inner-city bodegas. Those millions of people deserve a label they can trust, one they can understand and that is consistent over time. We must work together to create the best definition of what organic means: humanely raised and grass-fed animals, social justice and fair trade standards, and worker rights all make the label more credible. We can make it true. Remember, the government is us.

Our resistance to unification over the organic standards is what makes it easier for the chemical companies to succeed. The many labels consumers see today—grass-fed, humanely raised, free-range, locally grown—reflect important value choices, but they make it harder for people to understand what each choice means and why it matters. Chemical companies know that most people make their decisions based on price, so they put all their energy and resources into keeping the price of food artificially low. As a result, many people give up trying to figure out all the labels and just buy the cheapest foods, especially in tough times.

Big business and industry are not inherently bad or evil. There are bad businesses and bad people, but scale doesn't necessarily make them that way. Big business, when focused on creating positive change, can make big changes happen fast. Big businesses can make more good things available to people more widely, quickly, and affordably.

The same people who invented and mass-produced chemical fertilizers and pesticides also developed aspirin and many other chemical products that have improved our lives. We need to sort out what is good from what is toxic and make the right decisions for the future. Demonizing those who don't see the world our way is a waste of precious time.

Furthermore, whether you trust the government or not, government regulation is essential. Thriving, healthy, growing economies and communities are built on trust, and on faith in that trust. But it is crazy to think that trust, without regulation, will suffice. When transactions go beyond one-on-one hand-shaking agreements, which they must in order for us all to survive, regulation becomes essential. And when trust fails, as it has in the past few years, our whole economy fails.

The same is true for organics. Regulations are essential to keep people honest, because there are so many steps in the food chain before products make their way to our tables. Unless you live totally self-sufficiently, you will never be able to control every aspect of your food, so we must find ways to trust in others to do as we would.

I am not suggesting that the government or anyone else should tell us what to eat. No one is perfect. Especially not me. I don't eat organic foods all the time. I wish I could, but I work full-time at multiple jobs, volunteer on the boards of several nonprofits, and have three kids between the ages of 28 and 3. We eat pizza, takeout, even fast food. When we travel, I sometimes pack snacks, but I am just as likely to pull over for Popeyes red beans and rice. Eating organic all the time, as others have discovered, is hard if not impossible. My dream is that one day it won't be.

I also believe if people want to eat organic Twinkies, we should let them. There is too much judgment in this world already. Taste is personal and regional, and we don't need food police to govern it. I've seen that the more you try to control what people eat or what they do, the more likely they are to rebel. Our supermarkets and fast-food places can exist, and thrive, in an organic world. In fact, I think they have to. People eat what tastes good to them, not because they like or prefer the chemicals in the food. And people eat it because it's cheap, not because they wouldn't rather have a delightful organic meal at a fancy restaurant. We must make organic foods more widely available at reasonable costs to everyone, at every income level.

When I was a kid, I grew up in a house that never had a box of Pop-Tarts in it. I went to friends' homes and stared longingly at the box with the beautiful, colorful, iced pastries on it. It just didn't seem fair that some kids were allowed to have that stuff and I wasn't. I felt deprived. Today, I am thankful for companies like Nature's Path, which makes organic toaster pastries—with icing! We don't eat them often, but they are there if we really want one. And as a result, my kids don't feel deprived in the same way I did (although I am sure they feel deprived in some other way).

It's not too late to change—and get healthier and happier! People who eat organic foods reduce their pesticide intake by as much as 90 percent, according to a study from the University of Washington.[11] Further, research by University of Colorado neuroscientist Christopher Lowry, PhD, found that certain strains of soilborne bacteria not only stimulate the human immune system, but also boost serotonin levels in mice. Low levels of serotonin are tied to depression and drugs that inhibit its reuptake in the brain are used as antidepressants.[12] Perhaps putting your hands in healthy organic soil can prevent depression! If we all farmed and gardened the organic way, we may not need the antidepressant drugs that we are finding in our water supply (and that have dangerous side effects).

A lot of people won't agree with these statements—money might even be spent trying to negate my message, like those "clean" coal commercials or the ridiculous pro-high-fructose corn syrup commercials. Such campaigns create doubt, play on people's insecurities, and perpetuate falsehoods in favor of corporate profit. We have to be prepared for desperate attempts to salvage profits.

IMAGINE A NEW FUTURE

Think for a minute about what the world could look like if we all worked together and made it organic. Let's take the leap in our minds. How would we make this audacious goal of going organic happen? And what would the world look and feel like?

First, like on the current organic farms, there would be a transition period. We wouldn't need more land for farming, but we would need more farmers farming differently. Fortunately, a lot of young people who currently can't afford the land are interested in farming. And a lot of people who are out of work may not mind working on a farm. But we would need to create a new model for farm employment, since we shouldn't just rely on farm families, illegal immigrants, and migrant workers.

Perhaps a required period of farm service could be a part of every public school education so teenagers could learn where their food comes from. Summer vacations from school were originally intended to ensure kids were available to help their families with the farming during the busy growing season. That work gave children important physical activity, lessons in responsibility, an understanding and respect for food and where it comes from, and an appreciation of the benefits of hard work.

The farms themselves would need to become more diverse—what a few chemical farmers in Iowa called "Old McDonald" farms, like the children's song. Instead of endless rows of corn and soybeans, the farms would have animals, vegetables, chickens and eggs, and a variety of crops sowed in an integrated rotation system that produces much more food and fertility over time.

We could convert all those corn and soybean fields into

pasture to raise grass-fed cattle and hogs. The hogs would be freed from their tight cages, nutrients would be returned to the soil rather than festering in toxic vats, and we'd have cleaner, healthier, and happier meat to eat. Remember, organically managed pastures sequester a lot of carbon, and animals that eat their natural diets have fewer diseases. Turning corn and soybean fields into pastureland would reduce the supply of those commodities enough that the fewer farmers who still grew them would get a better price for their crops. And the meat produced would be much healthier and better tasting for all of us, not just those who can afford grass-fed diets now. Plus, instead of sitting in big tractors all day, becoming bored and overweight, farmers could get back in the saddle and become cowboys again.

Farmers will need help changing, since to most of them organic is a foreign way to farm. They are so used to relying on advice from chemical-company–funded consultants and universities that they'll need to get to know their land in a new way. They will need to relax their often rigid ideal of having totally "clean" and neat weed-free fields. Many farmers might even need emotional support. In the focus group I met with in Iowa, one older farmer was beginning to question his role in creating dead zones in the Gulf of Mexico. I could see fear and conflict in his face as he realized that maybe he hadn't been doing the right thing all his life after all.

This reeducation needs to start early—way before kids become farmers or consumers. The farm bill allocated more than $1 billion a year for the school lunch program. Instead of using that to feed kids leftover crap from farms and factories, we can use that money to get Edible Schoolyard organic gardens (or something similar) and curriculum into every public school. Kids can learn how to grow and make their own food, and as a result they will learn how to eat. Nothing is more likely to get kids to eat vegetables than to let them grow and pick them themselves.

Our whole landscape would change for the better. Cities would be surrounded by a wreath of small farms, greenhouses, and community gardens. Growing fruits and vegetables closest to where they are consumed just makes sense—food tastes best when it's fresh. Housing developments could become productive and fertile sources of food, like Prairie Crossing in Grayslake, Illinois. Edible landscaping would be part of how we all think. And it would taste good as well as look good!

Of course, the government would need to change, too.

In the business world, a farm-related reference is commonly used to describe an organization that is hierarchical, territorial, and, all too often, ineffective. The workers are said to be stuck in silos. On a farm, silos are used to store grains and silage (fermented plants that are fed to animals—but should really be made into compost). All government regulatory organizations are set up in silos, inhibiting cooperation and collaboration. We can't really afford that approach any more. And the last thing we need is another government bureaucracy to make producing food even more complicated and expensive. I urge President Obama and his successors to merge the EPA,

USDA, FDA, and maybe even the National Institutes of Health, Centers for Disease Control and Prevention and the Department of Energy into one single-integrated agency to create a strong, lean, and more effective system. An awful lot of money, time, and headaches could be saved and insights gained by having these groups work together, collaborating to find solutions that cross the divides.

If the chemical companies still want to stay in business, they can redirect their researchers to find ways of "harvesting" the plastic crap that they created and we have dumped into the oceans, and recycling it into something inert and useful.

How would the consumer experience change?

Supermarkets would still exist, thankfully. (I like supermarkets.) But they might be more like super-farmers'-markets, offering a combination of local and regional foods. The ridiculous segregation of organics that persists in our supermarkets—"hippie" zones—would go away. People could still get all their favorite products, but they would all be certified organic—even Doritos and Wonder bread.

Food prices might rise, but with heavy taxation and penalties on chemical companies, chemical foods would be much more expensive than organic. And of course, all chemical foods would have to have warning labels that list all the dangers of eating them, including death.

Health care costs, conversely, would decrease significantly. Slowly but surely, as we transitioned to organic, autism rates would decline; cancer rates would decline; diabetes rates would decline, obesity would decline (we'd all lose weight, thanks to removing all those hormone-disrupting, endocrine-disrupting chemicals from our lives!); allergy, asthma, and depression rates would go down. Infertility, miscarriages, genital deformities, and sexual dysfunction would decrease and babies would be born healthier.

People would start to feel better and be healthier. Ask any dairy farmer how things change after they go organic and they will tell you their cows are healthier and happier. Kids especially, with all the outdoor activity and farmwork, would become stronger and more vital. We'd see fewer behavioral problems and conditions such as ADHD.

There would be more frogs! More bees! More fireflies! More bats! (And therefore fewer mosquitoes!) Floods and droughts would be much less extreme and much less devastating.

While much would change, many things would feel the same. We would still have TVs, computers, and restaurants (though the food would taste a whole lot better). Our kids would still want to go shopping, only now all the cotton clothing they're begging for would be organic, which feels a lot softer against the skin (it's true!). There would still be fashion and fads. (Some things will never change.)

Doesn't that seem like a vision worth championing?

I know it's possible.

We all need to help and the first place to start is in our own lives and yards.

We need to plant more gardens. The beautiful thing about

organic and nature and mycorrhizal fungi is that they make it not really matter what you plant. Plant tomatoes or zinnias, herbs or a forest. Stake a claim, mark your ground, and take a stand. Make your commitment to staying on this planet and on this earth. Make it with your own personal sense of style and freedom of taste. Plant foods you love to eat, and use that organic garden to help you learn about nature, not fight against it. You'll find something unexpected will grow in your garden— trust. Trust in yourself. Trust in nature. Trust in our future, and finally, trust in our ability to make our future good.

8. FIVE SOLUTIONS THAT MIGHT SAVE US

You now know the problem we all face—we are suffocating and exterminating ourselves as the result of our tendencies to see nature as the enemy and allow chemical companies to trick us into believing we can control it. And I hope I have cleared up some of the confusion about organic farming methods and foods. Now, what can we all do about it?

We must restore the earth's natural ability to absorb and store carbon. Going organic will do not only that, but it will heal many other major ills as well: the poisoning of our children, our water, our wildlife, and our world.

There are five clear and fairly straightforward things that need to happen, and each sector of society has an important job to do.

1. GOVERNMENT: BAN AGRICULTURAL CHEMICALS AND GMOS

We need to demand that the government stop rewarding businesses that harm people and the planet by giving them subsidies and tax breaks and easing regulations. We need to demand that the government be on the side of the American people. We owe it to future generations to put aside our philosophical differences and partisan politics and, for once, unite for what is right for America and for the world.

■ **The government must stop subsidizing chemical farming.** We need to completely overhaul the farm bill to encourage as many farmers as possible to transition to organic as quickly as possible. We need to reorient the incentives so that the prices of organic foods and agricultural fibers reflect their real costs and are affordable.

■ **The tax structure should be aimed at taxing polluters at the highest rates and giving tax breaks to the people and companies that are doing the right things.** We need to reverse the model to incentivize businesses to do the right things. We need to levy major taxes and penalties for the contamination chemical companies have caused. When their products become more expensive—so consumers have to pay more to harm their health and destroy the environment—their use will decline. That idea is a lot closer to true free-market capitalism than what we now have. (Forget crazy cap-and-trade schemes, keep it simple!) Or, we could just ban them completely. That would be best.

■ **Now, more than ever, we need regulations that protect all of us.** We need fair trade regulations that protect people we will never see or meet, but who make things we use every day. We need health regulations that protect children around the world from chemicals that harm them. We need global agreements that enable us all to work together to solve the climate, energy, health, and food crises—next to them, the global financial crisis pales in comparison.

■ **We need to be global leaders (not babies or bullies!).** Whenever I hear politicians, pundits, or corporate leaders say that we won't make changes unless China and India agree to "go first," I am ashamed of our country. First, we created the problem. And even the poorest among our poor have it much better than the poor people of India and China. Second, and most important, *true leaders go first.*

■ **We need to keep the USDA organic standards pure.** Even now, lobbyists are working hard to weaken the organic rules so chemical companies can sell their toxic products to farmers and get the higher prices commanded by organic foods. Keeping the USDA Organic seal pure requires constant vigilance. It's time that we acknowledge these companies as the bullies that they are and have the courage and integrity to stand up to them to protect our organic standards and make them *better.* Without the unifying focus of organic, we will continue to run willy-nilly from cause to cause, trying to solve all of our problems piece by piece. They can't be solved in isolation.

■ **We need to encourage all elected officials to make the tough calls to cut programs that aren't work-ing so we can spend money on things that do work—and pay off the national debt.** We need to acknowledge that when the government does not subsidize education, business interests will step in to fund research and education as

extensions of their marketing budgets. We shouldn't allow
the chemical companies to underwrite 4-H and Future
Farmers of America clubs. We shouldn't allow them to control
our land-grant research universities with their funding. We
shouldn't allow biotech companies to obstruct independent
research! We should not allow them to advertise to influence
children.

I don't mean to suggest that these changes will come easily
or without protest. But I take heart in a precedent for this sort
of change. At one time, tobacco companies ruled the world and
people smoked anywhere they wanted to—on planes, at work,
in restaurants, in schools. Doctors appeared in ads promoting
cigarette brands. When my first daughter was born in 1982, my
hospital roommate smoked in the presence of our newborn
babies! Now, although cigarettes remain on the market, it is
almost impossible to smoke in a public place. We don't have the
50 years it took to gradually wrest control from the tobacco
companies. (It took 70 years for the research that showed
smoking caused cancer to be taken seriously, a period over
which millions and millions of people died.)

This time, we—and especially those in government—need
to be bold, decisive, and undeterrable.

2. FARMERS:
SUPPLY THE ORGANIC DEMAND!

As any addict knows, there is no halfway. You are either hooked
and at the mercy of your drug, or you are free and clear. Gaining

your freedom from the chemical companies might hurt for a little while, but what freedom isn't worth fighting for? Pick up your shovels and pitchforks and fight for your freedom! Demand that the government fund and support your transition to organic. Demand that chemical companies stop intimidating you and controlling your life.

We need all farmers to go organic, to break their addiction to the false lure of getting higher profits with synthetic-chemical farming. Every single farmer, large or small, local or international, needs to change. If American farmers don't make the switch, we will be importing even more of our food from other countries.

No one gets filthy rich farming, whether they use chemicals or organic methods, are industrial-scale or family farms, large or small (they might have gotten really rich 50 or 100 years ago, but those days are gone). But you can get rich in a different way—and experience the *true* feeling of wealth that comes from knowing you are doing something good. As corny as it sounds, at the end of the day, it's all about love: Love for the land and family, love for the life, love for nature and heritage and tradition. Love for food. Love for the children you want to see grow and be healthy and inherit a healthy, love-filled world.

I know farmers can find the courage to step off the treadmill that is leading them to nowhere they really want to go. I know farmers can find the strength to stand up for what is right for the land, their families, and their futures. I know they can and will.

3. BUSINESS:
CREATE INNOVATIVE SOLUTIONS

The business community needs to support and enrich people and the planet—not the other way around. There is a movement growing around the idea that businesses can be leaders in making positive change for the world. It started with the concept of social responsibility and now has names like conscious capitalism and natural capitalism. Many successful companies now reflect this fresh modern ethic—companies such as Patagonia, Stonyfield Farm, Nature's Path, Organic Valley and, yes, Whole Foods.

John Mackey, CEO and cofounder of Whole Foods, is a firm believer in the power of business to change the world. (I know he changed my world for the better.) Mackey believes that the notion that capitalism is evil and business is the bad guy comes from an outdated view of the world in which everything is a "morality play that divides the world up into good and evil." More than anything else, he believes (as do I) that we must believe we can create "positive solutions for the way the world can evolve." And when we don't like something, we must "criticize by creating."

While businesses—or, more accurately, the people who run them—have done some pretty bad things over the centuries, they've also done a lot of good. We need to sort out what we want to keep and what works from what we don't need anymore and what's holding us back.

This is an exciting time for entrepreneurs—there are so

many challenges in need of creative and innovative solutions. It's a great opportunity for inventors, investors, and idea people to find good things that work well and make them available to people on a wide scale. That's business at its best and most exciting.

If you happen to work for a chemical or biotech company, know that you have the power to stop it from within. You can learn and decide to do the right thing. You can refuse to do the chemical company's bidding. If you are a researcher at a university, you can refuse to take their money. You can stand up for what's right. You can stop being part of an industry that is destroying the species—including your own children. You can use your skills, talents, expertise, and efforts to heal the planet. It's up to you.

4. ECONOMISTS:
MEASURE STRENGTH, NOT GROWTH

At the root of our global economic crisis and our agricultural mistakes is the belief that we can and must grow at all times, that growth is the single most important measurement. When, inevitably, growth slows down or stops, everyone panics, and fear makes it all worse. At its very root, our economic crisis is an emotional crisis, with fear and panic ruling our behaviors.

The so-called laws of economics are not only a fairly recent inventions, they are also deeply flawed. They are the result of our outdated industrial paradigm, which leads us to believe that if we have enough fuel, we can keep the factory running forever and people will just keep consuming.

But if we look at economics through an organic paradigm, we quickly see that nature is in a constant state of change—of growth and recovery, night and day, action and rest. The stock market behaves in the exact same way, but we look at it and see failure instead of natural growth and recovery. In the organic paradigm, recessions are kind of like Sundays, days off to enjoy our families and restore our strength for the next week of work. Or, like winters, which, as long as we are expecting it and prepared for it, can be a season we enjoy (or at least know it will end at a fairly predictable time). If we expect and even look forward to the seasons of an organic economy, it will minimize the "fear and panic" that can escalate recessions into depressions.

To use a natural metaphor, some trees grow very fast, but they tend to be fragile, short-lived, and prone to sudden death. Others are very slow growing and long-lived, surviving for hundreds and even thousands of years. But all trees have periods of growth (spring) and periods of dormancy (winter)—even tropical trees. All trees have value over both the short and long term. Seeing the growth of a slow-growing tree is harder, but it often provides more value for a longer period of time.

The true crime of Wall Street and our economic model is that they instilled the expectation and valuation of constant growth rather than the recognition of the cycles inherent in nature and the valuation of long-term strength.

5. EVERYONE: DEMAND ORGANIC

Organic is something we can all partake of and benefit from.

When we demand organic, we are demanding poison-free food. We are demanding clean air. We are demanding pure, fresh water. We are demanding soil that is free to do its job and seeds that are free of toxins. We are demanding that our children be protected from harm.

We all need to bite the bullet and do what needs to be done—buy organic whenever we can, insist on organic, fight for organic and work to make it the norm. We must make organic the conventional choice and not the exception available only to the rich and educated.

All of us have to stand together and stand strong for food that is safe and affordable for all. Alone, we are tiny ripples in a pond. Together, we are the waves that can turn stone into sand. Together, we *can* create a delicious and healthy future for everyone. We still have a choice of the future that will be ours, but we need to make our choice quickly.

I've made my choice.

We are all afraid of change—especially when we can't see the outcome. In business and life, I often call it the leap of faith. You may not know exactly where you will land, but you have to believe you will land somewhere good. Sometimes you just have to jump.

A major barrier to change is the hard time we have making connections—understanding, for instance, that soil and how we grow our food matter in ways that personally affect us.

Usually the phrase "everything is connected" sounds mystical. It's easy to discredit mystics and say they are abnormal, ungrounded, or crazy. But in nature, everything truly is connected, including us, because we are part of nature, too. It's like one of those hidden-picture puzzles—it's there, we just have to train our eyes and minds to see it that way. There's nothing mystical about it. Today, I can e-mail people in China or call them on the phone and get an immediate response. We are all connected, and we have no excuse for remaining ignorant about those connections and what they mean to us and our future.

Nature creates everything. Nature gives everything. Nature takes everything away. Nature feeds us, clothes us, shelters us. We, as part of nature, can invent cool stuff from nature. When we destroy nature, we are destroying ourselves. Nature is beyond faith, beyond religion, beyond politics, beyond race. Nature just *is*, and we just are part of it, too. We don't have to figure out how to control it; we just have to learn how to work together.

Doing nothing is not an option anymore. We can be a beacon of hope for the rest of the world. But we need the whole world to join us in our reformation. Because we are all in this together.

Ultimately, we are all one tribe.

We have a choice. Which future will we choose? We can't afford incremental change. We need transformative change.

We need an organic revolution. Our survival depends on it.

This is my organic manifesto.

EPILOGUE

I believe it is time for a new human experiment. The old experiment
... is that we have sprayed pesticides, which are inherent poisons ...
throughout our shared environment. They are now in amniotic fluid.
They're in our blood. They're in our urine. They're in our exhaled
breath. They are in mothers' milk.... What is the burden of cancer
that we can attribute to this use of poisons in our agricultural
system?... We won't really know the answer until we do the other
experiment, which is to take the poisons out of our food chain,
embrace a different kind of agriculture, and see what happens.

—Sandra Steingraber, Ithaca College, Scholar in Residence, Department of Environmental Studies and Sciences, President's Cancer Panel Report

The challenging thing about writing a book like *Organic Manifesto* is that the story never ends. Twists and turns in the plot appear daily. But since I turned in the original manuscript, many very interesting things have happened that further underscore the urgency of my message.

Shortly before my book was first published in March 2010, a team of scientists at the University of Illinois published research showing that artificial chemical nitrogen fertilizer contributes to global warming in previously unrecognized ways.[1] They also found that long-term usage of this fertilizer destroys the productivity of soil.[2] Ironically, this idea is not new at all—Sir Albert Howard, author of *An Agricultural Testament,* wrote about the

danger of soil depletion from artificial nitrogen use in 1940. One
of the authors of the study, Dr. Richard Mulvaney, who holds a
PhD in soil fertility and chemistry from the University of Illinois
and is now a professor in the university's Department of Natural
Resources and Environmental Sciences, does credit Sir Albert
Howard despite the fact that this historical information has been
long overlooked in the curriculum in that school, which receives
generous bio-tech industry funding. Not coincidentally, the
University of Illinois is home to the Morrow plots, where corn
growth and productivity have been studied since 1876.

Reporting on the study in *Grist*, Tom Phillpott noted:

> A particularly stark set of graphs traces soil organic
> carbon (SOC) in the surface layer of soil in the Morrow
> plots from 1904 to 2005. SOC rises steadily over the first
> several decades, when the fields were fertilized with
> livestock manure. After 1967, when synthetic nitrogen
> became the fertilizer of choice, SOC steadily drops.[3]

This is important stuff, but it may not register with the same
immediacy as the findings of the *2008–2009 President's Cancer
Panel: Reducing Environmental Cancer Risk—What We Can Do Now*,
which came out in the spring of 2010. Written by two George W.
Bush–appointed doctors, the panel reported that the risks of
artificial chemicals in our environment are understudied and
most likely responsible for much more cancer than we currently
realize. A careful reading of the report shows that few researchers are trained to look for environmental or human health
hazards of chemicals and that, as the study reports, the industry

has exploited the lack of research and government oversight in order to release untested chemicals into the enviroment.

The panel also reports that children who grow up on chemical farms have consistently higher rates of leukemia than other children, as do those who live in houses where pesticides and insecticides are used regularly.

Additionally, the report states that the US standards for safety are significantly more lax than international standards, a state of affairs that impacts not only our health as a nation, but also our ability to engage in global trade.

With each new finding, every frightening new revelation, I ask myself: What is wrong with us? How can we allow these loosely regulated industries to poison us and our children? Would we truly not be willing to pay a little more for real, healthy, organic food and keep our children healthy?

And that's not the end of the bad news.

In May, a study conducted at the University of Adelaide, Australia, ranked all countries in the world on their environmental degradation.[4] Only Brazil bested the United States for the distinction of "most degraded environment." No doubt had these rankings been compiled just a few months later,[5] after an oil rig in the Gulf of Mexico exploded, unleashing an unprecedented gusher of oil for four months, the United States would have taken the top spot. Larry Schweiger, president of the National Wildlife Federation, spent many weeks in the Gulf region after the April 20, 2010, spill and is gravely concerned not only with the impact of the gusher itself but also with the long-term effects of the chemical dispersants used to "hide" the oil. He says, "The BP well explosion, in addition to killing workers, polluted the Gulf with 4.1 million barrels of oil, 2 billion barrels (equivalent) of methane and 1.5 million gallons of chemical flocculants. Seven hundred thousand gallons of the chemical Corexit were pumped into waters that were one mile deep.

"Much of the oil and Corexit went to the bottom. This untested approach turned a more manageable two-dimensional oil spill into a more complex three-dimensional one, as oil dragged to the bottom breaks down slowly in dark, cold waters with low oxygen and little sunlight. This practice has never been tested and was outside of the chemical registration required by EPA. Patents and trade secrets protect the exact chemical makeup of Corexit yet all of this oil dispersant has been fed to the fish and shellfish of the Gulf with unknown consequences. Anyone who suggests they know the extent, duration or magnitude of the fisheries and ecological damages caused by this spill and the poor response efforts is either grossly misinformed or they are not telling the truth."

Speaking of the Gulf of Mexico, 2010 was the worst year so far for its ever-growing dead zone.[6] Unrelated to the oil spill, agricultural farm runoff deposited into the Gulf by the Mississippi River has now created an oxygen-deprived area the size of Massachusetts in which nothing can survive— and this is in *addition* to the area contaminated by oil and dispersants.

There's more.

■   According to NASA, 2010 was the fourth warmest year
on record.[7] And CO_2 levels are continuing to increase at a
frightening pace.[8]

■   Nashville flooded, likely as a result of soil degraded from
chemical farming and no longer able to store water as it should.

■   In a fairly predictable attempt to revive the tarnished
reputation of high-fructose corn syrup (HFCS), the corn
industry is trying to re-brand HFCS as "corn sugar."

■   Oh, and despite research that shows Roundup (glyphosate) causes birth defects[9] and weed resistance, 40 percent of
farmers still plan on increasing their use of glyphosphate[10] and
mixing it with other, more potent herbicides to fight ever stronger weeds. Monsanto is even paying for farmers to use other,
more toxic herbicides to augment the effectiveness of Roundup.

And I'm just starting to understand enough about fracking
to recognize that it's bad. Really bad. This process involves
using lots of water to remove natural gas from shale, and in
that process, drinking water can become so polluted that
neither people nor animals should drink it.

Is there any good news? A bit.

The pesticide Aldicarb was banned—an interesting story
for a number of reasons. First, this comes twenty-five years
after it caused the worst pesticide poisoning in American
history. And second, that event inadvertently illustrated that
the idea that chemicals can be "peeled off or washed off" was a

lie. How so? Well, in the summer of 1985, 2,000 people got sick from eating watermelon contaminated with Aldicarb (made by our friends at Bayer Crop Sciences).[11] They didn't eat the rind of the watermelon; they ate the flesh inside, yet they were still sickened by the presence of the chemical in the fruit. As this example shows, chemicals are systemic. You *cannot* always avoid them by washing or peeling your produce.

(Note that even though Aldicarb has been banned, it won't be phased out entirely until 2015. Until then you might want to avoid potatoes, citrus, watermelon, and cotton that are not certified organic, since these are the primary chemical crops that are treated with Aldicarb today.)

And in one of the few studies comparing chemical food to organic, the organic strawberries won![12] Yes, they were smaller since they weren't doped up on chemicals, but they had more antioxidants and lasted longer without rotting, and the soil they grew in was much more alive with microbes (that's a good thing for storing carbon!).

Finally, Monsanto, despite being named Forbes' Company of the Year for 2009 (disgusting!), had fairly poor financial results in 2010. It may be too soon to hope that the GMO tide has turned, but for the first time in many years, farmers are planting fewer of Monsanto's GMO seeds, not more. Monsanto's GMO sugar beets got turned down by the government—for now—and farmers are starting to complain about the high price of GMO corn seed that does not perform as promoted.

Furthermore, new research from the Rodale Institute is showing that while Monsanto has set a stretch goal of increasing the drought tolerance of their GMO seeds by 6 percent to 13 percent, organically farmed crops are already 31 percent more drought tolerant.[13] Before we celebrate the demise of GMO seeds, though, it should be noted that competitor Pioneer's GMO seeds are picking up market share from Monsanto.

When will we learn? What will it take to finally create the change we need?

In a private meeting this year with a junior policy aide to an experienced senator, I was told the problem in Washington isn't so much the money that Monsanto gives to candidates, it's the pressure—the PR pressure, the legal and lobbying pressure—that the company, as a campaign contributor, puts on politicians if they show any sort of support for organic. Of course, all that pressure takes lots of cash—cash harvested from farmers who believe they need chemicals they don't really need, from government subsidies, and from people who are too willing to just buy whatever food is cheapest and easiest. The only antidote to this sort of pressure is the pressure that we citizens en masse can put on our elected officials, which thankfully also seems to be happening.

I am angrier now than when I first started writing *Organic Manifesto*. But I can't create the change that is needed alone, and I need all your help to figure out what to do next. Where is our courage? By the time you read this, much more will have changed, some for the better and some for the worse. But one thing won't change, and that is the fact that each of us bears personal responsibility for our world. I need each and every one of you to demand positive, thoughtful, and constructive change. Spread the word, and for the sake of all of us, DEMAND ORGANIC!

ACKNOWLEDGMENTS

There is quite a distance to be traveled between getting an idea for a book and bringing it to completion. That distance can't be measured in miles or hours or even drafts of manuscripts (of which there were many). For me, the most accurate measure is the knowledge gained and mental expansion that occurred from when I started to when I finished, and the interactions with all the wonderful people who helped me along the way.

I'd like to start by thanking my Dad. Even though he has been dead for 20 years, if he hadn't dragged us all around the country (and to Europe once or twice) to visit farms and farmers, attend conferences and meet interesting people, and if he hadn't dragged home all sorts of dignitaries to our dinner table (much to my mother's chagrin), I'm not sure I could have developed my passion and commitment to the organic movement quite so deeply and so early. I don't think it's a coincidence that the idea for this book came to me on his birthday.

Among the living, I'd like to thank Diana Erney, research librarian at Rodale Inc., who from that first lunch in the South Mountain Building cafeteria—when she jumped on board with long pent-up fervor—has guided and grounded me all along the way. Only she and I will ever know just how inappropriate, scandalous, and angry those first few drafts of the book were.

She had the courage to disagree with me, the confidence to point out weaknesses in my arguments, and the patience to track down obscure facts that were critical to the credibility of this book.

A big thanks goes to Dan Weise, who met me in three different states, on the outskirts of towns, and brought me not only the most interesting groups of farmers, but also gave me the benefit of his lifetime of insights into chemical farmers and the world they live in (often shattering stereotypes I had in my head) and conveying the nostalgic longing farmers have for the times when farming was a simpler way of life connected to strong families and communities. Of course I also have to thank all the farmers Dan found; they really opened up to one another, sharing their trials, tribulations, frustrations, and fears—and allowed me to listen.

And to all people who graciously allowed me to interview them, including Deborah Koons Garcia, Myra Goldman, John Mackey, and all the people at the Rodale Institute, especially Alison Grantham, George Bird, Paul Hepperly, and Tim LaSalle. I am especially grateful to Warren Porter and Dr. Phil Landrigan.

To the people in Washington who have aided the Institute over the years and been wonderful supporters—Congressman Charlie Dent (R), Senator Bob Casey Jr. (D), and Senator Arlen Spector (R/D). And to Pennsylvania representative David Kessler, who is so courageous.

And to Mal Gross, who did his best to keep me from getting into trouble. Lord knows, many have tried.

For their insightful editing and guidance, I'd like to thank Pam Krauss and Karen Rinaldi, who strengthened my narrative arc and patiently put up with my sudden change in job in the middle of the book—with unexpected visits from both author and boss. You handled it well! And to both Andy Carpenter and Amy King for their design work on the book jacket, and to Joanna Williams who did the interior design. Thanks also to Nancy Bailey for production editing—and what the heck, Reds, too!—who both happen to be some of my favorite people on the planet.

I must thank ALL the people at Rodale who kept the business going when I had to focus on writing. Their commitment to the company during these most challenging times is inspiring and heart-warming, and their patience over the past few years has been much appreciated. I'd specifically like to thank Bernadette Eckhart and Mark Kintzel, my team for keeping my life in order—well, as much as they can, they try. Rick Chillot, Leah Zerbe, Emily Main—my fantastic editors at Rodale.com are always uncovering great stories just when I need them most, and David Kang made sure I saw every important article because he knew I needed to see them. And thanks to Robin Shallow and all the PR and marketing people at Rodale, who work their hearts out for every author, including me.

Speaking of our authors, to the Honorable Vice President Al Gore, for all his work on exposing global warming and finding solutions, and for his faith in Rodale. He is an inspiration to many, but especially to me. And to Al Gore's smart and wonderful special projects manager Brad Hall, who read the book and gave me critical feedback and links just in the nick of time.

There are many people who along the way offered me encouragement, support, and advice: Steve Murphy, Leigh

Haber, George DeVault, Scott Meyer, Ethne Clarke, David Fenton, John Grogan, all the farmers at the Emmaus Farmers' Market, and many others. Thank you!

For their organic leadership over the years I'd like to thank Eric Schlosser, who graciously agreed to write the foreword to this book, Wendell Berry, Joan Gussow, Gary Hirshberg, Barbara Kingsolver, Marion Nesle, Michael Pollan, Anthony and Florence Rodale, Bob Scowcroft, Vandana Shiva, and Alice Waters; I am hoping we can unite together to demand organic! And to Bette Midler and Drew Becher for planting trees and showing that healing the planet can be fun. I am especially grateful to Irving Dardik, who showed me the truth about understanding nature. If it wasn't for Scott Schultz, from Shultz and Williams, I wouldn't have the phrase "demand organic," which is the perfect rallying cry for this book. Thanks Scott!

To country music—for making me love farming even more, and helping me see the world through different eyes— and realize that inside we are all pretty darn much the same. (Best farming songs include: *Big Green Tractor* and *Amarillo Sky* by Jason Aldean, *She Thinks My Tractor's Sexy*, by Kenny Chesney, *International Harvester* by Craig Morgan and heck— anything by Keith Urban and Sugarland, even if it has nothing to do with farming.)

Essential thanks go to my family: Maya, for reading many versions and for joining the Rodale Institute (and Tony, for joining our family); Eve, for asking great questions and for helping out around the house (and with Lucia); Lucia, for keeping me laughing; and last but certainly not least, Lou, for seriously holding down the fort while I went through the craziest, most stressful and challenging time of my life so far (and that's a whole other book!).

Finally, I would like to honor the memory of my mother, who passed away as this book was going to press. She was one of the fiercest defenders of organic I ever met—and yet until the end still refused to switch some of her favorite brands. From her I learned just how hard change can be, but that it *is* possible!

NOTES

INTRODUCTION

1. I use the phrase "grown with synthetic chemicals" to mean all farming that uses man-made chemicals such as fertilizers, herbicides, fungicides, and pesticides to manage the production of crops.
2. Organic agriculture, as defined by the USDA, is food grown without the use of man-made chemicals, without seeds that have been genetically modified, without the use of irradiation or sewage sludge, and, for animals, without hormones or antibiotics.

CHAPTER 1

1. Andrew C. Martel, "Cases of Lead-Tainted Wells Climb in North Whitehall," *Morning Call,* December 26, 2008, B5; Martel, "EPA Finds Elevated Arsenic, Lead Levels," *Morning Call,* March 3, 2009, B1; Martel, "EPA Studies Options for Removing Lead-, Arsenic-Tainted Soil," *Morning Call,* March 11, 2009, B1.
2. John D. Meeker et al., "Cadmium, Lead, and Other Metals in Relation to Semen Quality: Human Evidence for Molybdenum as a Male Reproductive Toxicant," *Environmental Health Perspectives* 116 (2008): 1473–79.
3. PediatricAsthma.org, "The Burden of Children's Asthma: What Asthma Costs Nationally, Locally, and Personally," http://www.pediatricasthma.org/about/asthma_burden.
4. American Academy of Allergy and Asthma Immunology, "Asthma Statistics," http://www.aaaai.org/media/statistics/asthma-statistics.asp.
5. Ibid.
6. NASA Earth Observatory, "The Carbon Cycle," http://earthobservatory.nasa.gov/Features/CarbonCycle.
7. The Royal Society of Medicine Health Encyclopedia. London: Bloomsbury Publishing Ltd, 2000. s.v. "carbon dioxide," http://www.credoreference.com/entry/rsmhealth/carbon_dioxide.
8. Biochar is a fine-grained, highly porous form of charcoal created by a process in which plant and animal wastes are burned. Because the process bypasses normal decomposition and also acts as a fertilizer, biochar is useful in sequestering soil carbon. International Biochar Initiative, "What Is Biochar?" http://www.biochar-international.org/biochar.
9. "Cap and trade" is a term used to describe emissions trading, in which a regulatory authority sets a limit, or "cap," on the amount of carbon a company can release into the environment in a single year. The company is given a certain number of credits annually, and if it doesn't use them by the end of the year, it can sell those credits on the open market. Companies that exceed their limits must purchase credits from companies that have produced fewer emissions. S. George Philander, ed., *Encyclopedia of Global Warming and Climate Change: Volume 1* (London: Sage, 2008), 364.

10. Novecta, "Charting a New Direction for Agriculture," http://www.novecta.com.
11. Carlin Flora, "Cult of Clean," *Psychology Today,* September/October 2008, 93–99.
12. United States Geological Survey, "Water Science for Schools: Irrigation Water Use," http://ga.water.usgs.gov/edu/wuir.html.
13. Robert J. Diaz and Rutger Rosenberg, "Spreading Dead Zones and Consequences for Marine Ecosystems," *Science* 321 (August 2008): 928–29.
14. Ibid.
15. Courtney D. Kozul et al., "Low-Dose Arsenic Compromises the Immune Response to Influenza A Infection *in Vivo*," *Environmental Health Perspectives* 117 (2009): 1441–47.
16. American Academy of Allergy and Asthma Immunology, "Asthma Statistics." See note 4.
17. Agricultural Health Study, "AHS Scientists Begin Study of Lung Health," *Iowa Study Update 2008,* http://aghealth.nci.nih.gov/pdfs/IAStudyUpdate2008.pdf.

CHAPTER 2

1. Centers for Disease Control and Prevention, "Autism Spectrum Disorders (ASDs)," http://www.cdc.gov/ncbddd/autism/data.html.
2. Liz Szabo, "Food Allergies in Kids Soar," *USA Today,* October 23, 2008, 7D.
3. Alice Park, "The Year in Medicine 2008: America's Health Checkup," *Time,* December 1, 2008, 41–51.
4. Global News Services, "USDA Halts Pesticide Testing," *Morning Call,* October 1, 2008, A8.
5. Mount Sinai, "Children's Environmental Health Center: Environmental Toxins," http://www.mountsinai.org/Patient%20Care/Service%20Areas/Children/Procedures%20and%20Health%20Care%20Services/CEHC%20Home/Environmental%20Toxins.
6. O. P. Soldin et al., "Pediatric Acute Lymphoblastic Leukemia and Exposure to Pesticides," *Therapeutic Drug Monitoring* 31 (2009): 495–501.
7. Newswire, "FDA Supports Ban on Antibiotic Use for Growth Promotion, Feed Efficiency in Animals," States News Service, July 19, 2009.
8. Iman Naseri, Robert C. Jerris, and Steven E. Sobol, "Nationwide Trends in Pediatric *Staphylococcus aureus* Head and Neck Infections," *Archives of Otolaryngology—Head and Neck Surgery* 135 (2009): 14–16.
9. Shuaihua Pu, Feifei Han, and Beilei Ge, "Isolation and Characterization of Methicillin-Resistant *Staphylococcus aureus* Strains from Louisiana Retail Meats," *Applied and Environmental Microbiology* 75 (2009): 265–67.
10. Margaret Mellon, "Testimony Before the House Committee on Rules on the Preservation of Antibiotics for Medical Treatment Act H.R. 1549," July 13, 2009, http://www.ucsusa.org/assets/documents/food_and_agriculture/july-2009-pamta-testimony.pdf.
11. Stephanie Woodard, "The Superbug in Your Supermarket," *Prevention,* August 2009, 102–109.
12. "EPA Completes Reregistration of Controversial Antibacterial Triclosan," *Pesticides and You* 28 (2008–2009): 4.
13. Evanthia Diamanti-Kandarakis et al., "Endocrine-Disrupting Chemicals: An Endocrine Society Scientific Statement," *Endocrine Reviews* 30 (2009): 293–342, http://www.endo-society.org/journals/ScientificStatements/upload/EDC_Scientific_Statement.pdf.

14. Harvey Karp, "Cracking the Autism Riddle: Toxic Chemicals, A Serious Suspect in the Autism Outbreak," *Huffington Post*, June 30, 2009, http://www.huffingtonpost.com/harvey-karp/cracking-the-autism-riddl_b_221202.html.
15. Melonie Heron et al., "Deaths: Final Data for 2006," *National Vital Statistics Reports* 57, no. 14 (2009).
16. Devra Davis, *The Secret History of the War on Cancer* (New York: Basic Books, 2007), 4.
17. Agricultural Health Study, "Important Findings from the Agricultural Health Study," http://aghealth.nci.nih.gov/results.html.
18. Agricultural Health Study, "Pesticides May Increase the Risk of Diabetes," *Iowa Study Update 2008*, http://aghealth.nci.nih.gov/pdfs/IAStudyUpdate2008.pdf.
19. T. Edward Nickens, "Who Turned Out the Lights: Firefly Populations Appear to Be Dwindling. The Question Is Why," *Garden and Gun*, August/September 2009, 30–31.
20. Diamanti-Kandarakis, " Endocrine-Disrupting Chemicals."
21. Matthew D. Anway et al., "Epigenetic Transgenerational Actions of Endocrine Disruptors and Male Fertility," *Science* 308 (2005): 1466–69.
22. Leon John Olson et al., "Aldicarb Immunomodulation in Mice: An Inverse Dose-Response to Parts per Billion Levels in Drinking Water," *Archives of Environmental Contamination and Toxicology* 16 (1987): 433–39.
23. Ibid.
24. Wade V. Welshons et al., "Large Effects from Small Exposures: I. Mechanisms for Endocrine-Disrupting Chemicals with Estrogenic Activity" *Environmental Health Perspectives* 111 (2003): 994–1006.
25. US Environmental Protection Agency, "2,4-D RED Facts," June 30, 2005, http://www.epa.gov/oppsrrd1/REDs/factsheets/24d_fs.htm.
26. Tyrone B. Hayes et al., "Characterization of Atrazine-Induced Gonadal Malformations in African Clawed Frogs (*Xenopus laevis*) and Comparisons with Effects of an Androgen Antagonist (Cyproterone Acetate) and Exogenous Estrogen (17beta-Estradiol): Support for the Demasculinization/Feminization Hypothesis," *Environmental Health Perspectives* 114 Suppl 1 (2006): 134–41.
27. US Environmental Protection Agency, "Decision Documents for Atrazine," April 6, 2006, http://www.epa.gov/oppsrrd1/REDs/atrazine_combined_docs.pdf.
28. Natural Resources Defense Council, "EPA Making Illegal, Secret Agreements with Pesticide Makers, Threatening Public Health, Lawsuit Charges," news release, February 17, 2005, http://www.nrdc.org/media/pressreleases/050217.asp.
29. Mae Wu et al., "Atrazine: Poisoning the Well: How the EPA Is Ignoring Atrazine Contamination in the Central United States," August 2009, http://www.nrdc.org/health/atrazine.
30. US Environmental Protection Agency, "Atrazine Updates," November 23, 2009, http://www.epa.gov/pesticides/reregistration/atrazine/atrazine_update.htm.
31. Wu et al., "Atrazine: Poisoning the Well."
32. A genetically modified organism (GMO) is an organism whose genetic characteristics have been altered by the insertion of a modified gene or a gene from another organism using the techniques of genetic engineering. *The American Heritage Medical Dictionary* (Boston: Houghton Mifflin, 2007). s.v. "genetically modified organism."
33. Monsanto, "Company History," http://www.monsanto.com/who_we_are/history.asp.
34. USDA Economic Research Service, "Data Sets: Adoption of Genetically Engineered Crops in the U.S.," July 1, 2009, http://www.ers.usda.gov/data/biotechcrops.

35. US Environmental Protection Agency, "Consumer Factsheet on: Glyphosate," http://www.epa.gov/ogwdw/contaminants/dw_contamfs/glyphosa.html.

36. Caroline Cox, "Herbicide Factsheet: Glyphosate," *Journal of Pesticide Reform* 24 (Winter 2004): 10–15; Lance P. Walsh et al., "Roundup Inhibits Steroidogenesis by Disrupting Steroidogenic Acute Regulatory (StAR) Protein Expression," *Environmental Health Perspectives* 108 (2000): 769–76.

37. Beyond Pesticides, "Scientists Call for 'Inert' Ingredient Disclosure," Beyond Pesticides Daily News Blog, January 23, 2007, http://www.beyondpesticides. org/dailynewsblog/?p=11.

38. American Academy of Environmental Medicine, "Genetically Modified Foods," May 8, 2009, http://www.aaemonline.org/gmopost.html.

39. Brian Halweil, "Still No Free Lunch: Nutrient Levels in U.S. Food Supply Eroded by Pursuit of High Yields," Organic Center Critical Issue Report, September 2007, http://www.organic-center.org/reportfiles/Yield_Nutrient_Density_Final.pdf.

40. "The Year in Medicine 2008: Genetically Modified Foods: China Has the World Worried," *Time*, December 1, 2008, 60.

CHAPTER 3

1. Deborah Koons Garcia, director, *The Future of Food* (Mill Valley, CA: Lily Films, 2004).

2. Many people believe that GMOs are just modern versions of hybrid seeds, but that is not the case. Hybrid seeds are created by manually manipulating pollen from plants that might not otherwise mate. GMOs are created by using a special gun to manually insert fragments of one or more separate species' DNA into that of plant seeds.

3. Based upon estimates by the International Service for the Acquisition of Agri-Biotech Applications, "Global Status of Commercialized Biotech/GM Crops: 2007," ISAAA Brief 37-2007 Executive Summary, http://www.isaaa.org/Resources/publications/briefs/37/executivesummary/default.html.

4. Stated by an Iowa farmer in a focus group meeting.

5. David S. G. Thomas and Andrew Goudie, *The Dictionary of Physical Geography* (Malden, MA: Blackwell, 2000), s.v. "salinization."

6. Charles Benbrook, "Impacts of Genetically Engineered Crops on Pesticide Use: The First Thirteen Years," Organic Center Critical Issue Report, November 2009, http://www.organic-center.org/reportfiles/13Years20091126_FullReport.pdf.

7. Todd R. Callaway et al., "Diet, *Escherichia coli* 0157:H7, and Cattle: A Review after 10 Years," *Foodborne Pathogens and Disease*, April 15, 2009 (6): 67–80.

8. Carolyn Lochhead, "Crops, Ponds Destroyed in Quest for Food Safety," *San Francisco Chronicle*, July 13, 2009, A1.

9. Colin McClelland, "Farmer Loses Battle in Biotech Dispute," Associated Press, May 24, 2004.

10. Paul Elias, "Enforcing Single-Season Seeds, Monsanto Sues Farmers," Associated Press, January 13, 2005.

11. Olivia Judson, *Dr. Tatiana's Sex Advice to All Creation* (New York: Henry Holt, 2002), 1–3.

12. Donald L. Barlett and James B. Steel, "Monsanto's Harvest of Fear," *Vanity Fair*, May 2008, 156–70.

13. Comment heard in a farmer focus group meeting.

14. USDA Economic Research Service, "Table 4: Certified Organic Producers, Pasture, and Cropland," Data Sets: Organic Production, September 9, 2009, http://www.ers.usda.gov/Data/organic/Data/PastrCropbyState.xls.

15. Chris Kenning, "Kentucky Goes After 'Marijuana Belt' Growers," USA Today, September 30, 2007, http://www.usatoday.com/news/nation/2007-09-30-kentucky_N.htm.

16. Ashfaque Swapan, "We Are at a Watershed: Vandana Shiva," India West, October 29, 2008, B1.

17. An article in New Scientist presented the results of a study disputing the facts and figures on Indian farmer suicides. However, the study was conducted by the International Food Policy Research Institute (IFPRI), whose mission is to bring "better seeds" to the Third World. (By the way, "better seeds" is generally a euphemism for GMO seeds. At the Clinton Global Initiative meeting, I even heard Bill Gates use the term in explaining how he was going to spend his billions to help bring "better seeds" to developing countries.) IFPRI is supported by the Consultative Group on International Agricultural Research (http://cgiar.org), which is a major recipient of funding from the Syngenta Foundation for Sustainable Agriculture. Andy Coghlan, "GM Cotton in the Clear Over Farmer Suicides," New Scientist, November 6, 2008, 14.

18. Beverly D. McIntyre et al., eds., International Assessment of Agricultural Knowledge, Science and Technology for Development (IAASTD): Global Report (Washington, DC: Island Press, 2009).

19. Deborah Keith, "Why Walking Out Was Our Only Option," New Scientist, April 8, 2008, 17–8.

20. Ibid.

21. "Arvind's Organic Cotton Project Saves Farmers from Suicide," Economic Times, May 13, 2008.

CHAPTER 4

1. Alexander von Humboldt and Aimé Bonpland, Personal Narrative of Travels to the Equinoctial Regions of America, During the Years 1799–1804, Volume I (London: Henry G. Bohn, 1852).

2. Jared Diamond, Collapse: How Societies Choose to Fail or Succeed (New York: Viking, 2005).

3. Francis J. Peryea, "Historical Use of Lead Arsenate Insecticides, Resulting Soil Contamination and Implications for Soil Remediation," Proceedings of the 16th World Congress of Soil Sciences, Montpellier, France, August 20–26, 1998.

4. Will Allen, War on Bugs (White River Junction, VT: Chelsea Green, 2008), xxviii.

5. Ibid., 22.

6. Ibid., 35.

7. Justus Liebig, Organic Chemistry in Its Applications to Agriculture and Physiology (Cambridge, MA: John Owen, 1841), 149.

8. Allen, War on Bugs, 37.

9. Diarmuid Jeffreys, Hell's Cartel: IG Farben and the Making of Hitler's War Machine (New York: Metropolitan Books, 2008), 40.

10. Stanley J. Kunitz and Howard Haycraft, eds., American Authors, 1600–1900 (New York: H. W. Wilson, 1938).

11. Richard A. Wines, *Fertilizer in America: From Waste Recycling to Resource Exploitation* (Philadelphia: Temple University Press, 1985), 34.
12. Jimmy M. Skaggs, *The Great Guano Rush: Entrepreneurs and American Overseas Expansion* (New York: St. Martin's Press, 1994), 4.
13. Wines, *Fertilizer in America*, 35.
14. Skaggs, *Great Guano Rush*, 2.
15. Allen, *War on Bugs*, 27..
16. Skaggs, *Great Guano Rush*, 9.
17. Skaggs, *Great Guano Rush*, 14.
18. "Guano Islands Act," Travel and History, http://www.u-s-history.com/pages/h1047.html.
19. Allen, *War on Bugs*, 28.
20. Skaggs, *Great Guano Rush*, 160.
21. Wines, *Fertilizer in America*, 48)
22. George Vaughan Dyke, *John Lawes of Rothamsted: Pioneer of Science, Farming and Industry* (Harpenden: Hoos, 1993), 89.
23. Royal Society (Great Britain), "The Founders of the Rothamsted Agricultural Station: A Sketch of the Life and Work of Sir John Bennet Lawes, Bart., F.R.S., and Sir J. Henry Gilbert, F.R.S.," *Obituary Notices of the Royal Society of London*, http://www.archive.org/details/foundersofrotham00royarich.
24. Dyke, *John Lawes of Rothamsted*, 90.
25. Wines, *Fertilizer in America*, 86–87.
26. Allen, *War on Bugs*, xxv.
27. Beyond Pesticides, "USDA and EPA Pushing Coal Ash for Agriculture Despite Toxicity Uncertainty," Beyond Pesticides Daily News Blog, October 20, 2009, http://www.beyondpesticides.org/dailynewsblog/?p=2571.
28. E. I. du Pont de Nemours and Company, *This Is DuPont: The Story of Farm Chemicals* (Wilmington, DE: E. I. du Pont de Nemours, n.d.).
29. Jeffreys, *Hell's Cartel*, 44–46.
30. Ibid., 89.
31. Allen, *War on Bugs*, 52–55; Fred Aftalion, *A History of the International Chemical Industry*, Chemical Sciences in Society Series (Philadelphia: University of Pennsylvania Press, 1991), 252.
32. Allen, *War on Bugs*, 83.
33. Michael A. Kamrin, ed., *Pesticide Profiles: Toxicity, Environmental Impact, and Fate* (New York: Lewis, 1997); Syngenta, Syngenta Crop Protection, http://www.syngentacropprotection.com.
34. Allen, *War on Bugs*, 107.
35. Dietrich Stoltzenberg, "Fritz Haber: Chemist, Nobel Laureate, German, Jew," *Croatica Chemica Acta* 78 (2005): A17–19.
36. Peter Hayes, *Industry and Ideology: IG Farben in the Nazi Era* (Cambridge, UK: Cambridge University Press, 1987), 361.
37. Jeffreys, *Hell's Cartel*, 267.
38. Ibid., 278.
39. Daniel Charles, *Lords of the Harvest: Biotech, Big Money, and the Future of Food* (Cambridge, MA: Perseus, 2001), 8.
40. Peryea, "Historical Use of Lead Arsenate Insecticides."
41. "Arsenic in Chicken Feed May Pose Health Risks to Humans," ScienceDaily.com, April 10, 2007, http://www.sciencedaily.com/releases/2007/04/070409115746.htm.
42. Allen, *War on Bugs*, 175.
43. Ibid., 147.

44. Cargill, "Our History," February 11, 2009, http://cargill.com/company/history/index.jsp.
45. Charles Wheelan, *Naked Economics: Undressing the Dismal Science* (New York: W. W. Norton, 2003), 36.
46. "George Washington," *Ohio History Central,* July 1, 2005, http://www.ohiohistorycentral.org/entry.php?rec=4.

CHAPTER 5

1. Department of Plant and Soil Sciences, Mississippi State University, "Grants Received During 2006," *Plant and Soil Sciences* 2, no. 2 (2006): 5–6.
2. Andrew Pollack, "Crop Scientists Say Biotechnology Seed Companies Are Thwarting Research," *New York Times,* February 20, 2009, B3.
3. Notice the language in this article in my local paper describing the group comprised of Monsanto, Archer Daniels Midland, Deere and Company, and DuPont (the italics are mine): "Organizers of the newly formed Alliance for Abundant Food and Energy said Thursday they want to *change the debate* about biofuels. Their plan is to *convince consumers and politicians* that both goals can be met at once by increasing agricultural productivity." Christopher Leonard, "Biofuels Lobby Is Formed," *Morning Call,* July 25, 2008, E2.
4. Keystone Center, "Diverse Group Releases First-of-Its-Kind Report Measuring Agriculture Sustainability," news release, January 12, 2009.
5. David J. Morrow, "Rise, and Fall, of Life Sciences; Drugmakers Scramble to Unload Agricultural Units," *New York Times,* January 20, 2000.
6. Leah Zerbe, "Media Miss Major Global Warming Contributor," Rodale.com, November 18, 2008, http://www.rodale.com/food-and-global-warming.
7. I. R. Dohoo et al., "A Meta-Analysis Review of the Effects of Recombinant Bovine Somatotropin: 1. Methodology and Effects on Production," *Canadian Journal of Veterinary Research* 67 (2003): 241–51; and I. R. Dohoo et al., "A Meta-Analysis Review of the Effects of Recombinant Bovine Somatotropin: 2. Effects on Animal Health, Reproductive Performance, and Culling," *Canadian Journal of Veterinary Research* 67 (2003): 252–64.
8. Andrew Martin, "Fighting on a Battlefield the Size of a Milk Label," *New York Times,* March 9, 2008, B7.
9. Katie Zezima, "Organic Dairies Watch the Good Times Turn Bad," *New York Times,* May 28, 2009, A12.
10. Jeffreys, *Hell's Cartel,* 41.
11. E. I. du Pont de Nemours, *This Is DuPont,* 30.
12. Andrew F. Smith, ed., *The Oxford Encyclopedia of Food and Drink in America* (New York: Oxford University Press, 2004), 351.
13. Wheelan, *Naked Economics,* 224.
14. Alessandra Rizzo, "UN: World Hunger Reaches 1 Billion Mark," Associated Press, June 20, 2009.
15. Tyron Richardson, "The Land Barge Is Back," *Morning Call,* January 21, 2009, A1.
16. Comment heard in a farmer focus group meeting.
17. Al Gore, *Our Choice* (Emmaus, PA: Rodale, 2009), 121.
18. Alex Crippen, "Warren Buffett and the Perils of Swimming Naked," CNBC.com, August 6, 2007, http://www.cnbc.com/id/20147026.
19. USDA Economic Research Service, *History of Agricultural Price-Support and*

Adjustment Programs, 1933-84: Background for 1985 Farm Legislation,
Agriculture Information Bulletin Number 485, December 1984.
20.   Ibid.
21.   Ibid.
22.   Ibid.
23.   Ibid.
24.   Earl L. Butz, "Crisis or Challenge?" *Nation's Agriculture* 46, no. 6 (1971): 19.
25.   Vernon P. Grubinger, "Organic Vegetable Production and How It Relates to LISA," *HortScience* 27 (1992): 733–61.
26.   J. Heckman, "A History of Organic Farming: Transitions from Sir Albert Howard's War in the Soil to USDA National Organic Program," *Renewable Agriculture and Food Systems* 21 (2006): 143–50.
27.   "Agribusiness Works to Define Sustainable Agriculture," *Environmental Leader,* July 8, 2009, http://www.environmentalleader.com/2009/07/08/agribusiness-works-to-define-sustainable-agriculture.
28.   Philip Mattera, *USDA Inc: How Agribusiness Has Hijacked Regulatory Policy at the U.S. Department of Agriculture* (Washington, DC: Corporate Research Project of Good Jobs First, 2004).
29.   Bradley S. Klapper, "UN Expert Faults US, EU Biofuel Use in Food Crisis," AP Worldstream, September 10, 2008.
30.   Jeff Etchason, "Advancing Medical Technology and Declining Health" (presentation, Lehigh Valley Hospital Network, Allentown, PA, March 4, 2009).
31.   Deborah Koons Garcia, director, *The Future of Food* (Mill Valley, CA: Lily Films, 2004); Marie-Monique Robin, director, *The World According to Monsanto* (Arte France Video, 2008).
32.   "Green Shoots: No Matter How Bad Things Get, People Still Need to Eat," *Economist,* March 21, 2009, 390.
33.   "Court: US Can Block Mad Cow Testing," *Morning Call,* September 1, 2008, A10.
34.   Organic Farming Research Foundation, "About Organic," http://ofrf.org/resources/organicfaqs.html.
35.   Tara Parker-Pope, "For 3 Never-Easy Years, Every Bite Organic," *New York Times,* December 2, 2008, D5.
36.   Wheelan, *Naked Economics,* 52.

CHAPTER 6

1.   Lord Northbourne, *Look to the Land* (Hillsdale, NY: Sophia Perennis, 2003), 58.
2.   US acreage: "Data Sets: Organic Production," http://www.ers.usda.gov/Data/Organic; global acreage: Helga Willer, Minou Yussefi-Menzler, and Neil Sorensen, eds., *The World of Organic Agriculture: Statistics and Emerging Trends 2008* (Bonn, Germany: International Federation of Agriculture Movements and Frick, Switzerland: Research Institute of Organic Agriculture, 2008), 26.
3.   According to the USDA, the number of women farmers increased by 29 percent nationally between 2002 and 2007. Arlene Martinez and Scott Kraus, "More Women Drawn to Farming," *Morning Call,* March 15, 2009, B1.)
4.   A roller-crimper, or cover-crop roller, is a tractor extension that was invented by the farmers at the Rodale Institute. In one pass, the roller-crimper mats down the cover crop and a rear-mounted Monosem no-till seed drill plants the seed. For more information, see http://www.rodaleinstitute.org/introducing_a_cover_crop_roller.

5. "Composting is, in broadest terms, the biological reduction of organic wastes to humus. Whenever a plant or animal dies, its remains are attacked by soil microorganisms and larger soil fauna and are eventually reduced to an earthlike substance that forms a beneficial growing environment for plant roots. This process, repeated continuously in endless profusion and in every part of the world where plants grow, is part of the ever-recurring process that supports all plant life." Deborah L. Martin and Grace Gershuny, eds., *The Rodale Book of Composting* (Emmaus, PA: Rodale Press, 1992), 1.
6. A swale is a low-lying or shallow depression in the land.
7. Organic Farming Research Foundation, "Organic Provisions in the 2008 Farm Bill," May 20, 2008, http://ofrf.org/policy/federal_legislation/farm_bill/080520_update.pdf.
8. Northbourne, *Look to the Land,* 59.
9. J. I. Rodale, *Organic Merry-Go-Round* (Emmaus, PA: Rodale Books, 1954), 6–7.
10. Henk van den Berg, "Global Status of DDT and Its Alternatives for Use in Vector Control to Prevent Disease," *Environmental Health Perspectives* 117 (2009): 1656–63.
11. Heckman, "A History of Organic Farming."

CHAPTER 7

1. David Pimentel et al., "Environmental, Energetic, and Economic Comparisons of Organic and Conventional Farming Systems," *BioScience* 55 (2005): 573–82.
2. "Finding the Real Potential of No-Till Farming for Sequestering Carbon," ScienceDaily.com, May 7, 2008, http://www.sciencedaily.com/releases/2008/05/080506103032.htm.
3. Don Comis, "No Shortcuts in Checking Soil Health," *Agricultural Research* 55, no. 6 (2007): 4–5.
4. David R. Higgins and John P. Reganold, "No-Till: The Quiet Revolution," *Scientific American,* July 2008, 70–7.
5. This fact was confirmed by farmers who participated in our focus groups.
6. Chensheng Lu et al., "Organic Diets Significantly Lower Children's Dietary Exposure to Organophosphorus Pesticides," *Environmental Health Perspectives* 114 (2006): 260–3.
7. Kai Ryssdal, "Using Technology to Grow More Food," interview with Monsanto CEO Hugh Grant, *Marketplace,* America Public Media, August 20, 2008, http://marketplace.publicradio.org/display/web/2008/08/20/corner_office_grant.
8. Andrew Martin, "How Green Is My Orange?" *New York Times,* January 22, 2009, B1.
9. World Resources Institute, *World Resources 2005: The Wealth of the Poor: Managing Ecosystems to Fight Poverty,* September 2005, http://www.wri.org/publication/content/7957.
10. International Union for Conservation of Nature, "Wildlife Crisis Worse Than Economic Crisis—IUCN," news release, July 2, 2009, http://www.iucn.org/?3460/Wildlife-crisis-worse-than-economic-crisis—IUCN.
11. Chensheng Lu, et al., "Dietary intake and its contribution to longitudinal organophosphorus pesticide exposure in urban/suburban children," *Environmental Health Perspectives,* April 2008, 116(4):537–42; Chensheng Lu, et al. "Organic diets significantly lower children's dietary exposure to organophosphorus pesticides." *Environmental Health Perspectives,* February 2006, 114(2):260–63.

12.   Daniel A. Marano, "Nature's Bounty: Soil Salvation," *Psychology Today*, September/October 2008, 57.

EPILOGUE

1.   S. A. Khan, et al., "The Myth of Nitrogen Fertiliztion for Soil Carbon Sequestration," *Journal of Environmental Quality* 36 (2007):1821–1832.

2.   R. L. Mulvaney, et al., "Synthetic Nitrogen Fertilizers Depelete Soil Nitrogen:A Global Dilemma for Sustainable Cereal Production," *Journal of Environmental Quality* 38 (2009):2295–2314.

3.   Grist.org, "New Research: Synthetic Nitrogen Destorys Soil Carbon, Undermines Soil Health," www.grist.org/article/2010-02-23-new-research-synthetic-nitrogen-destroys-soil-carbon-undermines-"

4.   C. J. A. Bradshaw. X. Giam, and N. S. Sodhi, "Evaluating the Relative Environmental Impact of Countries." PLoS ONE 5(2010): e10440. doi:10.1371/journal. pone.0010440

5.   www.guardian.co.uk/environment/2010/jun/29/bp-oil-spill-timeline-deepwater-horizon

6.   Reuters, "Gulf of Mexico Dead Zone Overlaps BP Spill Zone." www.reuters.com/article/idUSTRE6713YZ20100802

7.   NASA Global Climate Change, "How Warm Was This Summer?" http://climate. nasa.gov/news/index.cfm?FuseAction-ShowNews&NewsID=409

8.   NASA Global Climate Change, "Carbon Dioxide Concentration," http://climate.nasa.gov/keyIndicators/index.cfm Carbon Dioxide

9.   A. Paganelli, et al., "Glyphosate-Based Herbicides Produce Teratogenic Effects on Vertebrates by Impairing Retinoic Acid Signaling,"Chemical research in toxicology. 2010 Aug 9. [Epub ahead of print] Abstract of *Clinical Research in Toxicology* (Online 9 Aug. 2010; DOI: 10.1021/tx1001749).

10.   Croplife.com, "Poll: Farmers Taking Action Against Glyphosate Resistance," http://www.croplife.com/news/?storyid=2886

11.   Environmental Health News, "Insecticide to Be Banned—Three Decades after Tainted Melons Sickened 20,000 People," www.environmentalhealthnews.org/ehs/news/aldicarb-phaseout

12.   Karen Kaplan, "Organic Strawberries Are Better—In Some Ways—Researchers Say," *LA Times*, September 2, 2010

13.   Rodale Institute, "Drought Tolerance Critical, Say DuPont and Monsanto. We Couldn't Agree More." www.rodaleinstitute.org/20100827_drought-tolerance-critical

INDEX

A

ADHD, 18, 37, 182
Agent Orange, x, 26, 83
Agfa, 78
Agricultural Carbon Sequestration
Standard Committee, 9
Agricultural Health Study, 15, 23, 25
Agricultural Research Service (ARS),
160
Agricultural Testament, An, 150, 195
Agricultural workers, 15, 54, 148, 178
Agriculture, 150
Alar, 39, 84
Aldicarb, 28–29, 199–200
Allen, Will, 66, 75
Alliance for Abundant Food and Energy,
90–91
American Academy of Environmental
Medicine, 34
American Coal Ash Association, 75
American Farm Bureau, 50, 91, 109, 110
Antibacterial products, 22–23
Antibiotics
for animal treatment, 21, 22, 95, 100,
168
bacteria-created, 47–48
organic foods free of, 167
role of, in drug-resistant infections,
22, 23
Apples, pesticide-treated, 84
Archer Daniels Midland, 83, 134
Arsenic, 4, 14, 25, 77, 84–85
Asthma, 6, 15, 18, 160, 165, 182
Astra Zeneca, 92
Atrazine, 24, 25, 30, 31, 51
Attention-deficit hyperactivity disorder
(ADHD), 18, 37, 182
Autism, 18, 24, 37, 181

B

Balfour, Eve, 150
Baron-Cohen, Simon, 24
BASF, 76, 78, 81
Bayer, 78, 80, 81, 83
Bayer CropScience, 81, 91, 134, 200
Beneficial insects, for pest control, 146

Bhopal chemical disaster, 29
Biodiversity, organic farming
increasing, 171
Biofuels, 8, 55, 91, 99, 100–102, 113–14,
115, 116, 117, 148, 162
Biotechnology, 16, 83, 84, 88, 95
Bird, George, 82–83
Bisphenol A, 24, 28
Bosch, Carl, 76
Bovine growth hormone, 84
BP oil spill, 197–98
Brockmann, Miguel d'Escoto, 43
Bry, Lynn, 11
Buffett, Warren, 9, 103
Bush, George H. W., 124
Bush, George W., ix, 18–19, 124, 196
Businesses, making positive change,
189–90
Butz, Earl, 106, 156

C

Cancer
causes of, x, 4, 14, 19, 20, 23, 24–25,
28, 37, 54, 81–82, 85, 165, 187,
196
childhood, 20–21, 84, 197
suppression of research on, 24, 25
Cap-and-trade programs, 8, 129
Capitalism, 111, 185, 189
Carbon
soil organic, 196
sources of, 7, 56
stored in organic soil, 12, 111, 149,
157, 159, 166, 179, 184
Carbon dioxide
climate change from, 6–7, 12
dairy farms and, 162
stored by organic farming, 159
toxicity of, 7, 12
Carbon monoxide, 7
Carbon sequestration, 8–9, 10, 12, 111,
136, 159, 160, 162, 163, 173, 179
Cargill, 43, 83, 86, 91, 134
Carson, Rachel, 152
Carter, Jimmy, 107, 124
Cascadian Farm, 147